Tributes
Volume 41

Kreisel's Interests
On the Foundations of Logic and Mathematics

Volume 31
"Shut up," he explained. Essays in Honour of Peter K. Schotch
Gillman Payette, ed.

Volume 32
From Semantics to Dialectometry. Festschrift in Honour of John Nerbonne.
Martijn Wieling, Martin Kroon, Gertjan van Noord, and Gosse Bouma eds.

Volume 33
Logic and Computation. Essays in Honour of Amílcar Sernadas
Carlos Caleiro, Fransciso Dionísio, Paula Gouveia, Paulo Mateus and João Rasga, eds.

Volume 34
Models: Concepts, Theory, Logic, Reasoning, and Semantics. Essays Dedicated to Klaus-Dieter Schewe on the Occasion of his 60th Birthday
Atif Mashkoor, Qing Wang and Bernhrd Thalheim, eds.

Volume 35
Language, Evolution and Mind. Essays in Honour of Anne Reboul
Pierre Saint-Germier, ed.

Volume 36
Logic, Philosophy of Mathematics and their History.
Essays in Honor of W. W. Tait
Erich H. Reck, ed.

Volume 37
Argumentation-based Proofs of Endearment. Essays in Honor of Guillermo R. Simar on the Occasion of his 70th Birthday
Carlos I. Chesñevar, Marcelo A. Falappa, Eduardo Fermé, Alejandro J. García, Ana G. Maguitman, Diego C. Martínez, Maria Vanina Martinez, Ricardo O. Rodríguez, an Gerardo I. Simari, eds.

Volume 38
Logic, Intelligence and Artifices. Tributes to Tarcísio H. C. Pequeno
Jean-Yves Béziau, Francicleber Ferreira, Ana Teresa Martins and
Marcelino Pequeno, eds.

Volume 39
Word Recognition, Morphology and Lexical Reading. Essays in Honour of Cristina Burani
Simone Sulpizio, Laura Barca, Silvia Primativo and Lisa S. Arduino, eds

Volume 40
Natural Arguments. A Tribute to John Woods
Dov Gabbay, Lorenzo Magnani, Woosuk Park and Ahti-Veikko Pietarinen, eds.

Volume 41
On Kreisel's Interests. On the Foundations of Logic and Mathematics
Paul Weingartner and Hans-Peter Leeb, eds.

Tributes Series Editor
Dov Gabbay   dov.gabbay@kcl.ac.uk

Kreisel's Interests
On the Foundations of Logic and Mathematics

edited by

Paul Weingartner

Hans-Peter Leeb

© Individual authors and College Publications 2020. All rights reserved.

ISBN 978-1-84890-330-2

College Publications
Scientific Director: Dov Gabbay
Managing Director: Jane Spurr

http://www.collegepublications.co.uk

Cover design by Laraine Welch

All rights reserved. No part of this publication may be reproduced, stored in a retrieval system or transmitted in any form, or by any means, electronic, mechanical, photocopying, recording or otherwise without prior permission, in writing, from the publisher.

Georg Kreisel 2014

CONTENTS

Preface   vii

AKIHIRO KANAMORI
Kreisel and Wittgenstein   1

GÖRAN SUNDHOLM
Kreisel's Dictum   33

ULRICH KOHLENBACH
Local Formalizations in Nonlinear Analysis and Related Areas
and Proof-Theoretic Tameness   45

CHARLES PARSONS
Kreisel and Gödel   63

DANIEL ISAACSON
Georg Kreisel: Some Biographical Facts   87

KENNETH DERUS
Chitchat with the Devil: Kreisel's Letters, 2002–2015   121

Preface

The contributions to this volume are from participants of the international conference *Kreisel's Interests – On the Foundations of Logic and Mathematics*, which took place from 13 to 14 August 2018 at the University of Salzburg in Salzburg, Austria. It was organized by Paul Weingartner and Laurenz Hudetz of the Department of Philosophy.

The aim of the conference was to focus on Kreisel's main interests in the domain of logic and the foundations of mathematics. That this could only ever be an incomplete undertaking was clear from the very beginning. And yet, such an incompleteness fits very well with Kreisel's critical modesty concerning all types of "foundations". Although there are many scholars who work in this domain and would have been able to provide an interesting contribution focusing on "Kreisel's interests", it was the idea to select a relevant sample: only scholars who had been in scientific and/or personal correspondence with Kreisel for an extended period of time were invited. We therefore regret that Dana Scott was unfortunately not able to attend the conference and that Matthias Baaz and Mark van Atten, though having attended, were not able to submit their talks as a contribution.

We do not say that this is a volume in honour of Georg Kreisel since that would clearly be against his personal modesty. If he were still alive, he would ask: "Why are you troubling yourself with such an unimportant task?" However, we also know that he would not only read the contributions to this volume with interest but also comment on them and write critical remarks to the authors. For I (P.W.) remember that when I was collecting (at the *Institut für Wissenschaftstheorie* on *Mönchsberg* together with my assistant) his scientific articles and arranging them into what would eventually be a collection of 16 volumes, he said to me that I should not waste time on such things. Later, however, he also told me on several occasions how handy these volumes had been when writing a new article.

The papers given at the conference have been revised and partially extended, taking into account the discussions and scientific communication at the conference. We would like to thank all the participants who have contributed to this volume for taking over this additional, very valuable task.

Furthermore, we would like to express our gratitude to the *Georg Kreisel Institut für Wissenschaftstheorie* for sponsoring the conference. Our thanks also go to Rodrigo Leeb for writing the code to draw the diagram in Sundholm's contribution and insert Kreisel's visa in Isaacson's contribution. Last but not least, we would also like to thank Dov M. Gabbay and Jane Spurr of College Publications for their kind cooperation.

Salzburg, April 2020   Paul Weingartner and Hans-Peter Leeb

List of contributors and editors

Kenneth Derus: He is a former director of the Center for Combinatorial Mathematics, the dedicatee of piano works by Kaikhosru Sorabji and others, and the author of *Memories and their Objects*. He lives in retirement near Chicago. khderus@sbcglobal.net

Daniel Isaacson: Wolfson College, Oxford University, Linton Road, Oxford OX2 6U, United Kingdom, daniel.isaacson@philosophy.ox.ac.uk

Akihiro Kanamori: Department of Mathematics, Boston University, Boston, Massachusetts 02215, United States of America, aki@math.bu.edu

Ulrich Kohlenbach: Department of Mathematics, Technische Universität Darmstadt, Schlossgartenstraße 7, 64289 Darmstadt, Germany, kohlenbach@mathematik.tu-darmstadt.de

Hans-Peter Leeb Department of Philosophy, University of Salzburg, Franziskanergasse 1, 5020 Salzburg, Austria, hanspeter.leeb@aon.at

Charles Parsons Department of Philosophy, Harvard University, parsons2@fas.harvard.edu

Göran Sundholm: Department of Philosophy, Leyden University, goran.sundholm@gmail.com

Paul Weingartner: Department of Philosophy, University of Salzburg, Franziskanergasse 1, 5020 Salzburg, Austria, paul.weingartner@sbg.ac.at

1
Kreisel and Wittgenstein

AKIHIRO KANAMORI

Georg Kreisel (15 September 1923 – 1 March 2015) was a formidable mathematical logician during a formative period when the subject was becoming a sophisticated field at the crossing of mathematics and logic. Both with his technical sophistication for his time and his dialectical engagement with mandates, aspirations and goals, he inspired wide-ranging investigation in the metamathematics of constructivity, proof theory and generalized recursion theory. Kreisel's mathematics and interactions with colleagues and students have been memorably described in *Kreiseliana* ([Odifreddi, 1996]). At a different level of interpersonal conceptual interaction, Kreisel during his life time had extended engagement with two celebrated logicians, the mathematical Kurt Gödel and the philosophical Ludwig Wittgenstein. About Gödel, with modern mathematical logic palpably emanating from his work, Kreisel has reflected and written over a wide mathematical landscape. About Wittgenstein on the other hand, with an early personal connection established Kreisel would return as if with an anxiety of influence to their ways of thinking about logic and mathematics, ever in a sort of dialectic interplay. In what follows we draw this out through his published essays—and one letter—both to elicit aspects of influence in his own terms and to set out a picture of Kreisel's evolving thinking about logic and mathematics in comparative relief.[1]

As a conceit, we divide Kreisel's engagements with Wittgenstein into the "early", "middle", and "later" Kreisel, and account for each in successive sections. §1 has the "early" Kreisel directly interacting with Wittgenstein in the 1940s and initial work on constructive content of proofs. §2 has the "middle" Kreisel reviewing Wittgenstein's writings appearing in the 1950s. And §3 has the "later" Kreisel, returning in the 1970s and 1980s to Wittgenstein again, in the fullness of time and logical experience.

Throughout, we detected—or conceptualized—subtle forth-and-back phenomena, for which we adapt the Greek term "chiasmus", a figure of speech for a reverse return, as in the trivial "never let a kiss fool you or a fool kiss

---

[1] Most of the essays appear, varyingly updated and in translation, in the helpful collection [Kreisel, 1990a]. Our quotations, when in translation, are drawn from this collection.

you".[2] The meaning of this term will accrue to new depth through its use in this account to refer to broader and broader reversals.

1 Early Kreisel

At the intersection of generations, Kreisel as a young man had direct interactions with Wittgenstein in his last decade of life. Kreisel matriculated at Trinity College, Cambridge, where he received a B.A. in 1944 and an M.A. in 1947. In between, he was in war service as Experimental Officer for the British Admiralty 1943–46, and afterwards, he held an academic position at the University of Reading starting in 1949. According to Kreisel [1958b, p.157], "I knew Wittgenstein from 1942 to his death. We spent a lot of time together talking about the foundations of mathematics, at a stage when I had read nothing on it other than the usual *Schundliteratur*." Indeed, they again had regular conversations in 1942, when we can fairly surmise that the 18-year old Kreisel would have been impressionable and receptive about the foundations of mathematics. For 1943–45, however, their generally separate whereabouts would have precluded much engagement. During 1946–47, after the war, they had regular discussions on the philosophy of mathematics, although Wittgenstein had not written very much on the subject for two years.[3] At that time, Kreisel wrote his first paper in mathematical logic, [Kreisel, 1950]. From 1948 on, they would only have had intermittent contact, as Wittgenstein had resigned his professorship in 1947 and Kreisel took up his academic position at Reading in 1949. By the end of 1949, Kreisel had submitted for publication his [1951] and [1952a], the first papers on his "unwinding" of proofs. Wittgenstein was diagnosed with prostate cancer in 1949 and died in 1951. In what follows, we make what we can of the "early" Kreisel of the 1942 and 1946–47 conversations, our perception refracted through his published reminiscences.

Nearly half a century afterwards, Kreisel [1989a] provided "recollections and thoughts" about his 1942 conversations with Wittgenstein.[4] Early paragraphs typify the tone (p.131):

> I was eighteen when I got to know Wittgenstein in early 1942. Since my school days I had had those interests in foundations that force themselves on beginners when they read Euclid's *Elements* (which was then

---

[2] I owe the use of this term to my colleague Jeffrey Mehlman's in his remarkable [2010, §7].
[3] See [Monk, 1990, p.499].
[4] What we quote from [Kreisel, 1989a] is taken from the English translation in [Kreisel, 1990a, chap.9]. [Kreisel, 1978c] provides a shorter account of the 1942 engagements with Wittgenstein.

still done at school in England), or later when they are introduced to the differential calculus. I spoke with my 'supervisor', the mathematician Besicovitch. He sent me to a philosophy tutor in our College (Trinity), John Wisdom, at the time one of the few disciples of Wittgenstein. Wittgenstein was just then giving a seminar on the foundations of mathematics. I attended the meetings, but found the (often described and, for my taste, bad) theatre rather comic.

Quite soon Wittgenstein invited me for walks and conversations. This was not entirely odd, since in his (and my) eyes I had at least one advantage over the other participants in the seminar: I did not study philosophy. Be that as it may, in his company (*à deux*) I had what in current jargon is called an especially positive *Lebensgefühl*.

Kreisel soon went on (p.133): "One day Wittgenstein suggested that we take a look at Hardy's [*A Course of*] *Pure Mathematics* together. This introduction to differential and integral calculus was a classic at the time, and, at least in England, very highly regarded." Kreisel thence put the book in a mathematical and historical context, mentioning that Wittgenstein "had only distaste" for it—"something in the style, and perhaps also in the content, was liable to have got in the way"—and opining that the "foundational ideal" in Hardy was passé and to be supplanted by Bourbaki. Kreisel then recalled (p.136):

In the first few conversations about Hardy's book, Wittgenstein discussed everything thoroughly and memorably. The conversations were brisk and relaxed; never more than two proofs per conversation, never more than half an hour. Then one switched to another topic. After a few conversations the joint readings came to an end, even more informally than they had begun. It was, by then, clear that one could muddle through in the same manner.

As a matter of fact, Wittgenstein in his 1932–33 "Philosophy for Mathematicians" course had already read out passages from Hardy's book and worked through many examples.[5] What Kreisel writes coheres with Wittgenstein having made annotations in 1942 to his copy of the eighth, 1941 edition of Hardy's book.[6]

Just before these remarks, Kreisel had given a telling example from the conversations (p.135): "If $y = f(x)$ is (the equation of) a curve continuous in the interval $0 \leq x \leq 1$ and such that $f(0) < 0$ and $f(1) > 0$, then f intersects the x-axis. The job was to compute, from the proof (in Hardy) a point of

---

[5]Cf. [Wittgenstein, 1979].
[6]See [Floyd and Mühlhölzer, 2019] for accounts and interpretations of these annotations.

intersection." This of course is the Intermediate Value Theorem, the classical example of a "pure existence" assertion. In a footnote, Kreisel elaborates: "The proof runs as follows. If $f(\frac{1}{2}) = 0$, let $x_0 = \frac{1}{2}$. Otherwise, consider the interval $\frac{1}{2} \leq x \leq 1$ if $f(\frac{1}{2}) < 0$, and the interval $0 \leq x \leq \frac{1}{2}$ if $f(\frac{1}{2}) > 0$, and start again. This so-called bisection procedure determines an x_0 such that $f(x_0) = 0$."[7]

Kreisel mentioned "constructive content" and how "[...] in the conversations one looked for suitable additional data". He elaborated elsewhere [Kreisel, 1978c, p.79]:

> Wittgenstein wanted to regard this proof as a *first step*, and restrict it by saying: the proof only gives an applicable method when the relevant decision (whether $f(\frac{1}{2})$ is equal to, greater than, or less than 0) can be done effectively (e.g. if f is a polynomial with algebraic coefficients).
>
> I still find Wittgenstein's suggestion (of a certain restriction) agreeable: *satisfaisant pour l'esprit*. But it is certainly not useful (since the restriction is hardly ever satisfied). A variant ([Kreisel, 1952b]) is *much more useful*: it applies when the restriction is only approximately satisfied, i.e. when one is able to decide not necessarily at $x = \frac{1}{2}$ itself, but sufficiently close to it (e.g. in the case of recursive analytic functions on $[0, 1]$).

Wittgenstein's suggestion here—what Kreisel finds "agreeable"—is quite astute, resonant with the Intermediate Value Theorem not being intuitionistically admissible. There are continuous functions for which it is not intuitionistically possible to decide for their values whether they are equal to, greater than, or less than 0.

The forth-and-back in the quotation about Wittgenstein's agreeable suggestion and then its lack of usefulness is a local *chiasmus* of some significance. Kreisel is best known today, of course, for pioneering the study of the constructive content of proofs and the metamathematics of constructivity. In recollections ([Kreisel, 1989a, p.131]) "still exceptionally vivid, though perhaps rose-colored", he is emphasizing in self-presentation the constructive content. The 18-year old Kreisel may fairly be said to have been launched into his lifelong

---

[7]This proof is a binary version of the original [Cauchy, 1821, note III] proof of the Intermediate Value Theorem, and there is a historical resonance here. Being a pure existence assertion, the formulation and proof of the Intermediate Value Theorem by Cauchy and [Bolzano, 1817] was a significant juncture in the development of mathematical analysis. Their arguments would not be rigorous without a background theory of real numbers as later provided e.g. by [Dedekind, 1872]. The glossy Dedekind-cut proof found in Hardy (§101) is embedded in that theory, and Wittgenstein raised issues about the extensionalist point of view generally— cf. [Floyd and Mühlhölzer, 2019].

work by these early conversations with Wittgenstein. Kreisel subsequently wrote (p.136): "After the war I had a chance to go into mathematical logic in more detail; in particular, into consistency (WF) proofs. Instead of pursuing Hilbert's aim of eliminating dubious doubts about the usual methods of mathematics a more compelling application (better: interpretation) of those proofs occurred to me. Once again, the issue was a kind of constructive content; not, however, for items in some mathematical textbook, but for all derivations in some current formal systems." This was the direction of Kreisel's initial, and incisive, work in mathematical logic published in [Kreisel, 1950], of which more below.

On Wittgenstein's side, through 1942 he was actually working as a hospital dispensary porter in London toward the end of the Blitz, coming up to Cambridge on alternate weekends to deliver lectures on the foundations of mathematics (and presumably meeting with Kreisel then).[8] During this period, he penned remarks that would be compiled into Parts IV–VII of the *Remarks on the Foundations of Mathematics*.[9] Part V of the *Remarks* has an extensive discussion of non-constructive existence proofs and Dedekind cuts—Hardy's approach to the reals.

Kreisel (p.137) went on to write that Wittgenstein lent him a copy of *The Blue Book* at the beginning of summer 1942 and that he returned it by its end. *The Blue Book* was a text that Wittgenstein had dictated for his 1933–34 "Philosophy for Mathematicians" course and of which only a few copies were maintained. In *The Blue Book* Wittgenstein first brought forth the textures of meaning and language that would be elaborated in the *Philosophical Investigations*, like "language games" and their understanding through "training" toward the beginning and what to make of "I am in pain" with respect to the "I" at the end. Notably, in the face of this Kreisel only mentioned raising a "malaise" with Wittgenstein about his notion of "family resemblances of concepts". Invested in mathematics, Kreisel gave as an example the mathematical concept of group with its subcategories, mentioning a latter-day motto of his, "*relatively* few distinctions for *relatively broad* domains of experience". He could be said to have sidestepped Wittgenstein's main thrust, as exemplified by his example of "game", where various games have family resemblances but there is no property joining all instances, and the generality may be open-ended and evolving.

---

[8]Cf. [Monk, 1990, chap.21, esp. p.443], [Wittgenstein, 1993].

[9]Cf. [Monk, 1990, p.438]. The part numbers given for the *Remarks* are evidently for its second edition [Wittgenstein, 1978].

At this point, we record a passage from the [Monk, 1990] biography, a part of which has been passed along several times about Kreisel vis-à-vis Wittgenstein:

> In 1944—when Kreisel was still only twenty-one—Wittgenstein shocked Rhees by declaring Kreisel to be the most able philosopher he had ever met who was also a mathematician. 'More able than Ramsey?' Rhees asked. 'Ramsey?!' replied Wittgenstein. 'Ramsey was a mathematician!'

Wittgenstein was steadily drawn to mathematicians for conversation and intellectual stimulation. In the early 1940s, he would have found interaction with Kreisel in the next generation newly stimulating.

The post-war, 1946–47 conversations may have been extensive and far-ranging, but we can only make something of two published recollections of Kreisel. The first is about style, from [Kreisel, 1978b, n.2]:

> The matter of jargon, or style, came up often in my conversations with W (from 1942 to his death in 1951). For example, once after W had invited F.J. Dyson, who at the time [1946] had rooms in College next to W's, to discuss foundations. Dyson had said he did not wish to 'discuss' anything because *what* W had to say was not different from anything everybody was saying anyway, but he wanted to hear *how* W put it. W spoke to me of the occasion, agreeing very much with what Dyson had said, but finding Dyson's jargon a bit 'odd'. On another occasion, W said: Science is O.K.; if only it weren't so grey.

This resonates with what Kreisel wrote at the end of [1989a], that "The expository style (of Wittgenstein's conversations, where 'expository' would not apply to discussions) was at any rate for me much more successful", and "Wittgenstein's favorite quotation: *Le style, c'est l'homme*". Beginning with Wittgenstein's "distaste" for the style of Hardy's book, one can venture that the young Kreisel imbibed a sensibility to "style" so construed, this later seeping into his mathematical approach and writing.

The other recollection involves consistency proofs and the unprovability of consistency. As mentioned above, from [Kreisel, 1989a, p.136] one has "After the war I had a chance to go into mathematical logic in more detail; in particular, into consistency proofs", and at that time he had done the work to be published in [Kreisel, 1950]. From that publication, we can gather that he had begun by assimilating the 1939 *Grundlagen der Mathematik II* of Hilbert

Kreisel and Wittgenstein 7

and Bernays.[10] Kreisel wrote in [1983a, pp.300f] about what would have been from 1946–47:

> A few days after receiving several short, reasonable explanations of Gödel's incompleteness proofs Wittgenstein opined full of enthusiasm that Gödel must be an exceptionally original mathematician, since he deduced arithmetical theorems from such banal—meaning: metamathematical—properties like *WF* [consistency]. In Wittgenstein's opinion Gödel had discovered an absolutely new method of proof.
> [...]
> What he meant was that the metamathematical interpretation (made possible by the arithmetization of metamathematical concepts) makes the relevant arithmetical theorems immediately evident. This can be compared to the geometric interpretation of algebraic formulas, such as $ax^2 + ay^2 + bx + cy + d = 0$, from which it becomes obvious that two such equations cannot have more than two common roots (x,y), since two circles can intersect in at most two points.

There is ample evidence that Wittgenstein had already become aware of some of the ins and outs of Gödel's incompleteness theorem a decade earlier in 1937, when Turing's work came out.[11] What Kreisel is drawing attention to is Wittgenstein's apprehension of a "new method of proof", the metamathematical interpretation making the relevant arithmetical theorems "immediately evident".

Kreisel is known to have lectured on "Mathematical Logic" at the Moral Sciences Club on 27 February 1947, with Wittgenstein chairing.[12] The subject was presumably on the work to be published in [Kreisel, 1950].

Kreisel in that [1950] deftly provided "constructive content" to the Gödel incompleteness theorem, first exhibiting the sensitivity to recursiveness that would be a hallmark of his subsequent work. Drawing out recursive aspects of the Hilbert-Bernays 1939 *Grundlagen der Mathematik II* proof of Gödel's second incompleteness theorem, Kreisel established, in modern terms, that the Skolemized form of Gödel-Bernays set theory has no recursive model, exhibiting as a corollary a formula of first-order logic which has a model but no

---

[10]Kreisel elsewhere in [1987, p.395] wrote of "consistency proofs (which I had learnt in 1942 from Hilbert-Bernays Vol.2)". This may have been, especially in the sense of first acquaintance, but the tenor of various other recollections would suggest first full assimilation later.

[11]Cf. [Floyd, 2001]. *Remarks on the Foundations of Mathematics* [1956], Part I, drawn from 1937 manuscripts, has Wittgenstein ruminating over Gödel's proof of the incompleteness theorem.

[12]Cf. [Wittgenstein, 1993, p.355].

recursive model. Discussing at the end the definability of predicates through diagonalization, Kreisel provided the following telling, footnote 4:

> A great deal has been written since Poincaré on diagonal definitions occurring in a system of definitions. A very neat way of putting the point is due to Prof. Wittgenstein:
> Suppose we have a sequence of rules for writing down rows of 0 and 1, suppose the pth rule, the diagonal definition, say: write 0 at the nth place (of the pth row) if and only if the nth rule tells you to write 1 (at the nth place of the nth row); and write 1 if and only if the nth rule tells you to write 0. Then, for the pth place, the pth rule says: write nothing!
> Similarly, suppose the qth rule says: write at the nth place what the nth rule tells you to write at the nth place of the nth row. Then for the qth place, the qth rules says: write what you write!

Kreisel is acknowledging the rule-following versions of the Gödelian contrary, as well as the Turing direct, diagonalization arguments as given by Wittgenstein in conversation. As Kreisel moved forward with his "unwinding" [1951, 1952a] for constructive content of known proofs, this marks a closure point for the formative time of his direct engagement with Wittgenstein. Significantly, Kreisel will return to this footnote in the fullness of time and with a altered perspective, as will be discussed in §3.

On Wittgenstein's side, he with a change of aspect wrote in 1947 about Turing and rules ([Wittgenstein, 1980, §1096]):

> Turing's 'machines'. These machines are humans who calculate. And one might express what he says also in the form of games. And the interesting games would be such as brought one via certain rules to nonsensical instructions. I am thinking of games like the "racing game". One has received the order "Go on the same way" when this makes no sense, say because one has got into a circle. For that order makes sense only in certain positions. (Watson.)
> A variant of Cantor's diagonal proof:
> Let $N = F(k, n)$ be the form of the law for the development of decimal fractions. N is the nth decimal place of the kth development. The diagonal law then is $N = F(n,n) =$ Def $F'(n)$. To prove that $F'(n)$ cannot be one of the rules $F(k,n)$.
> Assume it is the 100th. Then the formation rule of $F'(1)$ runs $F(1,1)$, of $F'(2)$ $F(2,2)$ etc. But the rule for the formation of the 100th place of $F'(n)$ will run $F(100, 100)$; that is, it tells us only that the hundredth place is supposed to be equal to itself, and so for $n = 100$ it is *not* a rule.

> (I have namely always had the feeling that the Cantor proof did two things, while appearing to do only one.)
> The rule of the game runs "Do the same as [...]"—and in the special case it becomes "Do the same as you are doing".

This intensional, "rule" version of Turing's undecidability argument showing that the diagonal rule cannot be among the listed rules[13] corroborates Kreisel's footnote.

2 Middle Kreisel

In 1953, Wittgenstein's literary executors Elizabeth Anscombe and Rush Rhees published *Philosophical Investigations* [1953], what would become Wittgenstein's main legacy, out of manuscripts intended for publication. In 1956, the executors and G.H. von Wright published *Remarks on the Foundations of Mathematics* [1956], out of sporadic, working manuscripts from 1937–1944. And in 1958, Rush Rhees published *The Blue and Brown Books* [1958], two crafted texts from 1933–1935 sparsely circulated but never intended for publication. Kreisel, well into his career publishing five papers a year in mathematical logic and having met Gödel in Princeton, took it upon himself to provide extensive reviews of both the 1956 and 1958 publications. Let us proceed to this "middle" Kreisel with respect to Wittgenstein. Beyond our focus on Kreisel, it is of interest to take account of these reviews as part of the initial reception of Wittgenstein's works, especially in light of the considerable scholarship now attendant to this corpus.

Kreisel in his review [1958b] of the *Remarks on the Foundations of Mathematics* (RFM) took the compilation as presenting Wittgenstein's philosophy of mathematics, and contributed to setting a negative tone for its interpretation for quite some time. It is to be remembered, first of all, that RFM consists of unpolished, ruminating remarks never intended for publication and exhibit an evolution of thought and focus. Something of this as well as residual positivities for Kreisel were conveyed by him at the end of his review in a "Personal Note", which reveals an anxiety of influence:

> I knew Wittgenstein from 1942 to his death. We spent a lot of time together talking about the foundations of mathematics, at a stage when I had read nothing on it other than the usual *Schundliteratur*. I realise

---
[13][Floyd, 2012] calls this "Wittgenstein's diagonal argument" and analyzes it in great detail with respect to Turing's 1936 paper.

> now from this book that the topics raised were far from the center of his interest though he never let me suspect it.
>
> What remains to me of the agreeable illusions produced by the discussions of this period is, perhaps, this: every significant piece of mathematics has a solid mathematical core (p.142, 16), and if we look honestly we shall see it. That is why Hilbert-Bernays vol.II, and particularly Herbrand's theorem satisfied me: it separates out the combinatorial (quantifier-free) part of a proof (in predicate logic) which is specific to the particular case, from the 'logical' steps at the end. Certain interpretations of arithmetic and analysis have a similar appeal for me. I realise that there are other points of view, but for the branches of mathematics just mentioned, I still see the mathematical core in the combinatorial or constructive aspect of the proof.
>
> I did not enjoy reading the present book. Of course I do not know what I should have thought of it fifteen years ago; now it seems to be a surprisingly insignificant product of a sparkling mind.

Whether Kreisel was personally miffed or not, Wittgenstein scholarship has shown that Wittgenstein often did not discuss directly with students and others at the time what was at "the center of his interest". The "agreeable illusions" is chiasmatic, as Kreisel by this time had incisively pursued "the combinatorial (quantifier free) part of a proof" in [1951, 1952a] and moreover had shifted the focus of consistency proofs onto such parts in [1958a].

As to the concluding "insignificant product of a sparkling mind", this would become quoted, but evidently the "product" is the literary executors', concocted out of varying working manuscripts.

Kreisel begins his review by discussing Wittgenstein's "general philosophy" as a sophisticated empiricism sensitive to the ways of language. Kreisel considers Wittgenstein's starting point to be (p.138): "he is not prepared to use the notions of mathematical object and mathematical truth as tools in philosophy." But Kreisel does not consider as convincing Wittgenstein's arguments against them, and (p.137) "his reduction to rules of language". For Kreisel, "the real objection to these notions is that, at any rate as far as I know, there does not exist a single significant development in philosophy based on them." With this pragmatic pronouncement, he simply skirts the depths of Wittgenstein's grapplings in RFM with the objectivity of rule-following. Kreisel's only allusion to this is in a footnote (p.138):

> [...] it should be noted that Wittgenstein argues against a notion of mathematical object (presumably: substance), [...] but, at least in places [...] not against the objectivity of mathematics, through his recognition of formal facts.

Having ferreted this out of Wittgenstein, Kreisel himself would later become known for the dictum, "the objectivity of mathematics over the existence of mathematical objects".[14]

Kreisel next gets to Wittgenstein on proof. While a large part of RFM is devoted to aspects of proof, Kreisel here focuses on proof as related to theorem and, later in the review, on the equivalence of proofs (see below). Kreisel takes up as two themes that "A theorem is a rule of language and the proof tells us how to use the rule", and "The meaning of a theorem is determined only after the proof".[15] Kreisel discusses the various ways Wittgenstein approaches these themes at some length, but then deliberately reverses proof and theorem (p.140):

> Quite generally, it is simply not true that proof is primary and theorem derived, that only the proof determines the content of a theorem. In fact, Wittgenstein is wrong in saying that generally we change our way of looking at a theorem during the proof (p.122, 30), but equally often we change our way of looking at the proof as a result of restating the theorem; [...]

Kreisel will maintain this in his thinking as a *chiasmus*, elaborated with examples, but one can see it as a sort of surface reversal which can be subsumed into the greater depths of Wittgenstein's thinking.

First and foremost, Wittgenstein in RFM is seeing mathematics as a multifarious edifice of procedures and conceptual constructions, one for which proofs and methods of deduction as embedded in practice are crucial. Kreisel, in flattening the situation to a dichotomy between proof and theorem, and then shifting the weight back to theorem, eschews the complexity of interplay and moreover actually reinforces the importance of argument and construction. While Wittgenstein emphasizes how a proof accrues to the meaning of a theorem both by newly delineating its interplay of concepts and by providing procedural means for its further application, Kreisel emphasizes that (p.141) "*a theorem becomes an assertion about the actual structure of its own proof*"—which while focusing on theorem is in line with Wittgenstein's thinking. Kreisel's other way of shifting from proof to theorem is to emphasize that a proof yields new theorems, e.g. about structures.[16] Again, this is in accord

---

[14]For example, [Putnam, 1975, p.70]: "The question of realism, as Kreisel long ago put it, is the question of the objectivity of mathematics and not the question of the existence of mathematical objects."

[15]Kreisel (p.136) refers to RFM II §39 for the first and RFM II §31 and III §30 for the second.

[16] Even much later, [Kreisel, 1983a, p.297] supports this versus Wittgenstein though with an

with Wittgenstein's thinking, according to which a proof as procedure and becoming method is autonomous and would prove perforce various theorems.

Second, Kreisel continues from the above displayed passage with (p.140):

> [...] e.g. if we are accustomed to the principle of proof that the totality of all subsets of a set is itself a set, we may reject it when it is pointed out to us that it is only valid for the notion of a combinatorial set and not, e.g. for the notion of a set as a rule of construction.

Pursuant of this—or with it as an anticipation—Kreisel in a later, critical part of the review, on "Higher Mathematics", writes (p.153): "Wittgenstein says (p.58, 6) that it was the diagonal argument which gave sense to the assertion that the set of all sequences (of natural numbers) is not enumerable." After describing the diagonal argument and posing it as a "definition" of non-enumerability, Kreisel then wrote (p.153): "What is wrong here? Well, after all there was a paradox, Skolem's paradox, which puzzled people. The mistake is to think that the diagonal argument applies *only* to the set of all sequences [...]". Kreisel's allusion to Skolem's paradox, in purported line with the above displayed quote, is a local *chiasmus* in itself—about proof, theorem, and now the set of all sequences. Contrary to what Kreisel said about the diagonal argument being applicable in only one situation, Wittgenstein on the cited page had ruminated about "the diagonal procedure" in its various aspects, and wrote, rather, that "it gives sense to the mathematical proposition that the number so-and-so is different from all those of the system". A few pages earlier (pp.55f), he had discussed the diagonal procedure as a method, e.g. of transcending the algebraic numbers, and had expressed skepticism about the "idea" that the real numbers are not enumerable. Kreisel's simple gloss is seen to be overshadowed by Wittgenstein's wide-ranging remarks on the diagonal procedure as proof.[17]

---

oddly drawn example: "A caveman conjectures that $a^2 - b^2 = (a+b)(a-b)$ is valid for all even integers. Of course he is right. But the proof shows that the theorem has nothing to do with the distinctions of even and odd, integer or fraction. Therefore one formulates (the more general theorem for *arbitrary commutative rings*. This notion is determined by those few properties of the even integers which enter in the proof of $a^2 - b^2 = (a+b)(a-b)$. The more general theorem is more appropriate to the proof; in short: it is more meaningful."

[17]Notably, Kreisel in his next review [1960], to be discussed below, went to the extent of providing a "Correction" to the present review, allowing that Wittgenstein's (p.251) "remarks can be given a little more sense if an intensional notion of function (*rule* of calculation) is considered", and then giving three viable meanings of "enumeration". This resonates with how Wittgenstein was exploring the use of the diagonal argument and [Kreisel, 1950, n.4] as discussed at the end of §1; we will return to this at the end of §3, about a *mea culpa*.

The rest, and most, of the review concerns the "philosophy of mathematics". Kreisel had taken as Wittgenstein's conclusion in "general philosophy" (p.137):

> He regarded the traditional aims of philosophy, in particular of crude empiricism, as unattainable. He objected to a mathematical foundation of mathematics because the concepts used in the foundation are not sufficiently different from the (mathematical) concepts described (p.171, 13) and, he thought (p.177) that there are no mathematical solutions to his problems. He said the aim of a philosophy of mathematics should consist in a clarification of its grammar [...]

For Kreisel, (p.143) "I do not accept his conclusions since I do no think that they are fruitful for further research." Again a pragmatic pronouncement, and after rejecting on these grounds Wittgenstein's main thrust, the "clarification of its grammar" as a matter of mathematical activity, Kreisel proceeds, over several pages, to counter Wittgenstein's negativity about foundations with the fruitfulness of contemporary investigations of set theory and of constructivity. On the latter, Kreisel is discerning about the differences between intuitionism and finitism, and here he does take Wittgenstein as making contributions to finitist investigations.

Having cast light from his direction on foundations, Kreisel in the concluding pages of the review returns to proof—the focus of Wittgenstein's "foundational" concerns—as newly to be considered in the wider context. While Kreisel had earlier chiasmatically shifted the weight to the range of theorems that a proof can prove, he gets here to the range of proofs and Wittgenstein's interest in characterizing the equivalence of proofs and how they might be compared. Kreisel writes that Wittgenstein (p.151) "does attempt to find a characterization of a very general sort by basing a comparison of proofs on the application, or, as he puts it (p.155, 46) on what I can do with it." Though he finds limitations to this, Kreisel in support raises non-constructive existence proofs and "what we can do" with them—which is allusive to his own researches along these lines and their inspiration in his early conversations with Wittgenstein. Wittgenstein in RFM had ruminated over Gödel's proof of the incompleteness theorem, mainly about its ostensible play with truth, provability, and consistency. Taking his arguments as "wild", Kreisel strikes a positive path through Gödel's arithmetization-of-syntax argument, delineating that (p.154) "all that one needs of the concept of truth is $\mathcal{R}$ or $\neg\mathcal{R}$." Wittgenstein, generally, raised issues around consistency as a formal concept, with respect to proofs and contradictions. Kreisel insisted on the fruitfulness, writing (p.156):

"proofs of consistency and, more generally, of independence yield, perhaps, a better control over a calculus than anything else."

In his review [1958b] of RFM, aside from taking up objectivity vs. objects Kreisel mainly addressed what he regarded as challenges posed by the text concerning proof and foundations of mathematics as per meaning and knowledge. Variously flattening aspects, he set out contrasting viewpoints of proof vs. theorem, of the fruitfulness of foundational investigations, and even of specifics of the diagonal argument, the incompleteness theorem, and consistency. In this, he elaborated and promoted constructive aspects of proof.

Kreisel's review [1960] of *The Blue and Brown Books* can be seen as complementary, in that the text deals more centrally with language, and so what should be addressed is set in the seas of language rather than the precisification of mathematics. It will be remembered (cf. §1) that Wittgenstein lent Kreisel a copy of *The Blue Book* in the summer of 1942. The books first advance the method that would serve to buttress the mature *Philosophical Investigations*. In brief, Wittgenstein heralds the notion of a "language game" to shed light on the foundations of logic: the method utilizes simplified snapshots of portions of human language use to clarify meaning, understanding, and thinking. For concepts and categories, there is an exploration of the limits of reductive possibility, to be seen in the plasticities of language. For Kreisel, (p.240) "[...] quite natural developments of Wittgenstein's considerations may be formulated as a *reduction to the concrete*; for want of a better term I shall call it semi-behaviourism (with respect to mental acts) or semi-nomimalism (with respect to abstract objects)." This encapsulated interpretation is what Kreisel will discursively discuss in the review, and at the end of the review is a telling summary (p.251):

> As to content, the ideas of the book seem to be most relevant to the discipline which studies what is concrete (and whose exact delineation is yet to be evolved). On the positive side there are descriptions of little noticed phenomena (phenomenology) and reductions to concrete terms of many situations that are in the first place viewed abstractly. As described above a wider sense of 'reduction' is appropriate than is used in crude positivism or nominalism. This work shows convincingly a natural tendency of being unnecessarily abstract. On the negative side, we have Wittgenstein's theoretical positions; on analysis, there are seen to be cogent consequences of philosophical doctrines, which, roughly speaking, overestimate what can be done in concrete terms. Since the former seem to be easily refuted they are used in *reductio ad absurdum* arguments applied to the latter.

> As an introduction to the significant problems [of] traditional philosophy the books are deplorable.[2]
>
> ---
> [2]This is largely based on a personal reaction. I believe that early contact with Wittgenstein's outlook has hindered rather than helped me to establish a fruitful perspective on philosophy as a discipline in its own right, and not merely for example as methodology of highly developed sciences. [...]

The last sentence and its footnote are darker still than what Kreisel had written in that "Personal Note" at the end of the RFM review [1958b], quoted at the beginning of this section.

In the body of the review, Kreisel rounds out his contentions about Wittgenstein's "reduction to the concrete" with (p.240) "some illustrations taken from the philosophy of mathematics." At first, Kreisel is broadly affirmative about how Wittgenstein describes (p.241) "often surprisingly successfully, situations which are normally considered to involve just those mental acts and abstract objects which he eliminates." Kreisel relates this to how (p.242) "detailed investigations in the foundations of mathematics"—of which he writes tellingly in a footnote "My own in this direction have certainly been influenced by the view of Wittgenstein's work here described"—"have revealed a similar situation with respect to a nominalist (finitist, or, more generally, predicative) elimination of such abstract objects as the totality of natural numbers or of functions." Kreisel points out how for a wide class of proofs Herbrand's theorem provides "an elimination in a quite precise and natural sense" and similarly, "in a large part of analysis, quantification over all real numbers can be eliminated". Concluding about Wittgenstein's "practice of philosophy", (p.242) "Both his examples and the studies in the foundations of mathematics show clearly that *we have a general tendency to describe language* and, in particular, mathematical practice, *by means of concepts whose level of abstraction is higher than the minimum actually needed.*"

In the extended Remark following, Kreisel significantly pulls back by suggesting that what he had earlier written (p.243) "may be too logically biased and even altogether pragmatic." Instead, "we may look at these books, particularly *The Brown Book*, as a contribution to the study of *what is concrete, of what is (immediately) given.*" On this he brings in (p.243) "the theoretical question of the existence of sense-data" and Wittgenstein's "seeing X as Y". With the latter, Kreisel is astute enough to bring out something that would be central to Wittgenstein's later thinking, though by calling them "phenomenological studies" he diminishes their logical import. Kreisel pronounces (pp.243f): "though even in his later book *Philosophical Investigations* these phenomenological studies have not gone far enough to establish a discipline, the later work is incomparably better in this respect than the books under review."

Proceeding, Kreisel next considers Wittgenstein's "theoretical positions", which he takes to be (p.244):

> [...] (i) negative assertions on what cannot be said (or: is not), such as what is common or essential to those cases which he describes as families of concepts, (ii) assertions on what should be accepted as a decisive criterion (equality or difference in) meaning, such as the actual use of a term, (iii) the identification of metaphysical distinctions with grammatical ones.

Addressing (i), Kreisel takes Wittgenstein as objecting (p.244) "(a) generally, to the introduction of an (abstract) object common to all instances of a general term, (b) to the assumption that a general term always corresponds to a (single?—presumably: well-defined) property." Addressing these, Kreisel again resorts to mathematical illustrations. For (a), he points out that properties of rotations in the plane and multiplication of complex numbers can be commonly derived from the group axioms, and while there is a distinction in the two applications (p.245) "it's a distinction without a difference" and "the distinction is not vivid". For (b), Kreisel alludes to "mechanical procedure" à la Turing, and notes that (p.246) "It seems very natural that one is not instantaneously convinced of correct characterisations even if the arguments are good on reflection." Finally, as to what is essential to a concept, Kreisel points to the great deal of clarity gained "by the rather surprising discovery that relatively few abstract structures were essential to the proofs in the greater part of current mathematics."

By remaining in the concrete and curtailed formulations of mathematics, Kreisel is reducing away from Wittgenstein's main thrust in *The Blue Book* about the contexts and ostensible workings of language and meaning. Wittgenstein [1958, p.17], discussing "our craving for generality", pointed out "We are inclined to think that there must be something in common to all games, say, and this common property is the justification for applying the general term 'game' to the various games; whereas games form a *family* the members of which have family likenesses", these overlapping in various ways. In a different direction (p.18), of "the man who has learnt to understand a general term, say, the term 'leaf', [...] We say that he sees what is in common to all these leaves; and this is true if we mean what he can on being asked tell us certain features or properties which they have in common. But we are inclined to think that the general idea of a leaf is something like a visual image, but one which only contains what is common to all leaves", there being no such visual image. Finally, Wittgenstein somewhat anticipates the analytical and reductive approach that Kreisel is taking, with (p.18):

> Philosophers constantly see the method of science before their eyes, and are irresistibly tempted to ask and answer questions in the way science does. This tendency is the real source of metaphysics, and leads the philosopher into complete darkness. I want to say here that it can never be our job to reduce anything to anything, or to explain anything. Philosophy really *is* 'purely descriptive'.

Addressing (ii) of the penultimate displayed quote, Kreisel continues to take a reductive, scientific approach (p.247): "As far as *actual use* of words is concerned," it "may refer to the words spoken" or "it may also mean the *real* role of the word (as Wittgenstein puts it) undistorted by the vagaries of linguistic expression." It will become increasingly understood that Wittgenstein generally meant, rather, the use in a broad sense in our ordinary language. More attendant to the "real role", Kreisel opines "[...] in the cases of the elimination of abstract terms [...] there seems no doubt about the actual use [...] But in other cases the whole problem is thrown back to what is conceived as the real role"—on this referring to his discussion of "non-constructive" in the foundations of mathematics in his RFM review.

Addressing (iii), Kreisel first recalls that in his RFM review, he (p.247) "also questioned the value of the 'reduction' of metaphysics to grammar." Here, he refers to "syntactic" and "truth under the given interpretation" in mathematical logic, and opines, "I see no evidence that the grammatical distinctions which are to replace (problematic) metaphysical ones, are going to be described by means of less problematic concepts." This *is* a valid point, especially in answer to the temptation to take schematic formalization as elucidation of the large domains of truth and language. For Kreisel: "[...] the reference to grammar is deceptive for two additional reasons: First, [...] one does not usually consider such questions as 'what is a noun' in a theoretical way [...]. Second, while it is apt to speak of a grammatical role of a word in a language, the difficulty of formulating this seems to be of an entirely different order from school grammar [...]". Wittgenstein's use of the term "grammar" may indeed be deceptive at first, but it will become increasingly understood that he was taking it not as some sort of syntactic classification, but rather a tying of meaning to rules, of uses of general semantic types as these are correlated with syntactic categories in utterance and use.

Stepping back, one can fairly get the feeling that Kreisel in his review did not come to terms with Wittgenstein's frontal engagement with language and meaning. The *Philosophical Investigations* had come out in 1953, and the editor of *The Blue and Brown Books* had subtitled it *Preliminary Studies for the 'Philosophical Investigations'*. Nonetheless, Kreisel insisted on pursuing a path akin to the one taken in his RFM review, of making reductive logical pronouncements and alluding to logical-mathematical examples—managing, along the way, to make positive remarks about the elimination of abstract objects e.g. through Herbrand's theorem. In the large, Wittgenstein had begun to explore the seas of language, its waves to and fro, when reduction does not work to get at meaning.

3 Later Kreisel

In the fullness of time, after having pursued and stimulated avenues of research in constructive mathematics and proof theory and having had a substantive engagement with Gödel, Kreisel in several publications came again to engage with the words and ways of Wittgenstein. Latterly meditating on these in dialectical interplay with his own work and experience, Kreisel exhibited in style and tone a new, if commemorative, acknowledgement. This "later" Kreisel we pursue through his publications in chronological order, now to the further purpose of setting out his latter-day evolving thinking about logic and mathematics.

[Kreisel, 1976a], "Der unheilvolle Einbruch der Logik in die Mathematik", appeared among a collection of essays on Wittgenstein in honor of G.H. von Wright. The title is from *Remarks on the Foundations of Mathematics* IV 24, "The disastrous invasion of logic into mathematics". Kreisel takes this up as a theme of RFM—this in itself evincing a new positivity about that work—and proceeds to articulate his own thinking along these lines in light of contemporary developments.

Kreisel at the beginning cogently summarizes his line of thought (p.166):

> The aspects (of proofs and rules) which are regarded as basic in (1) current—somewhat pretentious—*logic*, are not only *different* from those which are essential in (2) current *mathematical practice* (which almost goes without saying), but actually *harmful* for a study of (2). The reason is that those basic questions of 'principle', concerning the validity of principles of proof and definition, appear more glamorous than the genuinely useful problems concerning current mathematical practice, and thereby divert attention from the latter. The 'practice' referred to in (2)

includes not only applications inside or outside mathematics, but also facts of experience concerning mathematical reasoning: which (combinatorial) configurations and (abstract) ideas we handle easily.

Then setting out toward elaboration, Kreisel instinctively retrenches (p.168): "[...] at least in my own case, the quotation has not been of direct, not even of heuristic use. I have known it for nearly 20 years, and stressed its plausibility in my—otherwise rather negative—review of RFM. The brutal fact is that the quotation does not contain the remotest hint of how (the pretentious) logical analysis is to be replaced, that is *which concepts should be used in the analysis of proofs in the place of the 'basic' concepts of proof theory* and which questions should be asked in place of the 'principal' problems of proof theory [...]". But later, "[...] the *value of Wittgenstein's quotation* (for me) can perhaps be summarized as follows: It is incisive and memorable, and so makes the reader familiar with a certain aim. If sometime later this aim is approximated, the reader is likely to take a closer look instead of moving on, breathlessly, to the next 'interesting' possibility."

Focusing on proofs and rules, Kreisel begins with the stark (p.169): "Proof theory is, in my opinion, a particularly crass example of that pretentious logic which was mentioned in the summary of this article [...] The claims of proof theory to have uncovered the true, in particular, formal nature of mathematical reasoning surpass in pretentiousness the claims of most traditional philosophers." This is a bit of *chiasmus*, a reversal toward Wittgenstein, in that Kreisel had himself proceeded in collaborative work in proof theory during this period with something of such "claims" as incentive.

Be that as it may, according to Kreisel, "Wittgenstein's critique of proof theory and its principal problems (for example in the *Remarks*) is wildly exaggerated, and therefore quite unconvincing." (p.170) "Worse still, Wittgenstein's own attempts to characterize what is essential in proofs aren't much better (than Hilbert's)." First, he "stresses that proofs *create*—or at least use!— *new concepts.*" Yet "the brutal fact remains that, somewhere or other, *propositions concerning these new concepts have to be proved too.*" And second, he stressed that "proofs must be graspable and memorable [...] and visualizable if we mean literal seeing of some spatio-temporal configuration [...]". "But all this is clearly secondary, as long as there are (genuine) doubts about the *principles of proof* that are used."

On this last, Kreisel makes an autobiographical remark revealing something of influence. He had a "long hesitation before studying the idea of simplicity or 'graspability' (Übersehrbarkeit) [sic] of proofs" (pp.173f):

> I just wasn't confident about finding a sensible measure in any direct way. First, I tried my hand at analyzing *simplicity* of principles of proof [...], by means of socalled autonomous progressions. Granted that these attempts were pretty faithful to the intended meaning, I soon came to this conviction: if the analyses are (even only) approximately right then those intended principles are just of little intrinsic interest [...] So instead I went back to more traditional questions about proofs, in particular, infinite proofs in intelligently chosen languages with infinitely long expressions, and, above all, intuitionistic logic. [...] What I overlooked was the witless way in which proofs entered! No recondite properties of proofs were involved, no relations between proofs or between proofs and other objects, nothing except their 'logical' aspects which occurs to us without any experience in mathematics at all! In short, nothing but the hackneyed business: The proofs establishes its conclusion (in particular a logic-free conclusion in the intuitionistic case).

Kreisel continued (p.174): "But, at last, I had become [...] convinced that questions of validity are by no means theoretically senseless [...] but that they are unrewarding at the present time." "At this stage it was natural to move so to speak to an opposite extreme, in particular, opposite to Hilbert's proof theory: I went about looking for methods of proof and properties of proof which are *trivial for proof theory*, but *essential for mathematical practice* [...] to be analyzed by appropriate mathematical measures of *complexity*."

On this, Kreisel gives two extended examples, the first being *explicit definitions* (p.174):[18]

> [...] we think of explicit definitions as *introducing* new concepts, the definition being usually supplemented by a list of properties (of the new concept), which are proved by the use of the explicit definition. As is well-known, this way of introducing a new concept is trivial for Hilbert's proof theory, because such concepts are in an obvious way *eliminable*. On the other hand, for mathematical practice they are not only useful, but as it were typical—at least for modern mathematics, which is dominated by the *axiomatic method*. This proceeds as follows. A structure is defined explicitly in set theoretic or number theoretic terms, and then is shown to be, say, a unitary group: the axioms for unitary groups then constitute the supplementary 'list of properties' (of the structure or concept) mentioned above. The choice of such properties—or, as one says, of the proper *cadre*—is often the key to solve mathematical problems.

---

[18][Kreisel, 1977] elaborates along some lines, and in particular has a longer subsuming account (pp.120ff) of explicit definitions.

For Kreisel, his student Richard Statman in his 1974 dissertation made (p.175) "impressive progress by means of a suitable measure of complexity which is *relevant in a large number of cases*, in particular, for analyzing the role of explicit definitions."

The second example (p.177) "concerns a more subtle 'invasion' by logic, namely a somewhat exaggerated idea of the role of socalled logical languages, for example, of predicate logic of first order", the exaggeration to be considered concerning "the ideal form of a (mathematical) proposition". On this, Kreisel focuses on real closed fields. After mentioning Sturm's work on determining the number of zeros of a polynomial in an interval and noting that *effective* decisions can be made when the coefficients are algebraic,[19] Kreisel thence brought in, of course, Tarski and the decidability of the first-order theory of real closed fields as a generalization. On this though, Kreisel opined (p.178) "The trouble began when people started to get interested in the efficiency of decision procedures [...]", and "[...] assumed that the 'ideal form' of 'the' decision problem for real closed ordered fields should deal with all formulas of the first-order language (of fields). They found so-called *upper and lower* bounds, namely $2^{2^{cn}}$ and 2^{cn} respectively, where n is the length of the formula." (p.179) "[...] the most obvious conclusion from the lower bounds is simply this: the full first-order language is *not* appropriate! And one would look for a *subclass* of that language (that has) a truly efficient procedure [...]". In the contemporaneous [1976b], Kreisel made proposals along these lines, and in [1982] worked out details for an application of Herbrand's Theorem for Σ_2 formulas.

After discussing related issues in budding computer science, Kreisel wraps up with "questions of 'principle' " and (p.186):

> I find it hard to have confidence in our whole 'critical' philosophical tradition, with it paradoxes, its dramatic claims either to see profound errors in our ordinary views or profound misconceptions in 2000 year old questions. It all sounds like a paranoid's paradise, and forgets the most striking fact of intellectual experience: how our thoughts seem to adapt themselves to the objects concerned, as we study them and get familiar with them (in a detached way) and how, with this familiarity comes the judgment need to distinguish between plausible and implausible theories, substantial and superficial contributions.

---

[19] Notably, this harkens back to that 1942 conversation with Wittgenstein (cf. §1) on "constructive content" of the Intermediate Value Theorem and Kreisel's effectivization in [Kreisel, 1952a].

After having elaborated in his own way about "the disastrous invasion of logic into mathematics", Kreisel here seems to come around to Wittgenstein's grounding faith in familiarity and the importance of our adaptability in coming to judgment.

[Kreisel, 1978b], "Lectures on the Foundations of Mathematics", is ostensibly a review of the compilation [Wittgenstein, 1976] of lectures notes for 1939 lectures put together by Cora Diamond. As if setting the stage, Kreisel quickly sketched Wittgenstein's progress *ab initio*, mentioning that (pp.98f) "W found that quite elementary mathematics provided excellent illustrations of weaknesses of traditional foundations, t.f. for short"—this incidentally setting a contrastive, positive tone from his RFM review.[20] But then, Kreisel shifts the purpose (p.99): "The main aim of this review is to restate the complaints of W and Bourbaki about t.f., with due regard for the discoveries of mathematical logic [...] By and large, at least in the reviewer's view, the discoveries of logic support the principal complaints." With this Kreisel virtually ignores [Wittgenstein, 1976], writing of it dismissively that (n.2) it "does not even record what W said in the lectures, but what a bunch of students thought he had said", and referring to it only on one page (p.107).

Kreisel takes the principal target of W and Bourbaki to be (p.99) "the *formal-deductive presentation* of mathematics in a *universal system*". But while "Bourbaki simply record their impression (of set-theoretic foundations)", Kreisel writes of Wittgenstein that (pp.99f) "[...] W attempts to convert fundamentalists by 'deflating' the notions and thus the so-called fundamental problems of t.f., stated in terms of those notions. In W's words, he wants to *show the fly the way out of the fly bottle*. He does this with much ingenuity and patience, and some overkill."

Proceeding to "complaints", Kreisel gets to (p.101): "[...] the general complaint (of W and Bourbaki) is that t.f. may be *poor philosophy*, in the broader popular sense of 'philosophy', specifically, if in practice the general aims of foundations are better served by alternatives, for example, by ordinary careful scientific research and exposition." Taking as "principal complaint: better current ideas than t.f.", Kreisel discusses how both W and Bourbaki emphasize that "the *choice of explicit definitions* is incomparably more significant than the glamorous preoccupations of t.f., not only for discovery, a 'mathematical' affair, but also for intelligibility, a principal factor in reliability."[21] Finally,

---

[20]Cf. §1.

[21][Kreisel, 1976a] elaborated on explicit definitions, as described in our account of it above. Significantly, Kreisel there wrote (p.175): "As far as I know, Wittgenstein himself never stressed the role of explicit definitions particularly." Now, he is accrediting to Wittgenstein how he

Kreisel addresses (p.102) "specific complaints about some glamor issues of t.f.". The first is "the matter of *contradictions* as in the paradoxes, or their absence, consistency, as in Hilbert's program." "W had a particularly strong aversion", whereas "at least by implication, Bourbaki was unimpressed". The second example is "higher (infinite) cardinals". As in his RFM review, Kreisel connects this to the diagonal construction, but now mentions favorably how Wittgenstein "preferred to use the construction in the context of rules", recalling Wittgenstein's formulation as given in [Kreisel, 1950, n.4].[22]

Throughout Kreisel's discussion of "complaints", there is in contrast to his RFM review a softer attitude toward Wittgenstein. This continues into Kreisel's acknowlegement (p.103) of "W's advice"—what mainly he draws from the book ostensibly under review—that "when confronted [...] by a philosophical problem about (mathematical) notions or proofs, we should see what we *do* with them, how we *use* them."[23] Kreisel concludes his "review" by "balancing the account on the positive side of t.f." He opined that to Wittgenstein the weaknesses of t.f. mattered less than the (pretentious) style, but proceeded to set out several examples—two from Gödel—for how such stylistic urgings may signal possibilities for progress.

[Kreisel, 1978a], "The motto of 'Philosophical Investigations' and the philosophy of proofs and rules," ostensibly takes up that motto, "All progress looks bigger than it is" interpreted as (p.13) "the ratio of *actual progress* (as judged by mature reflection) to *apparent progress* (measured by expectations after a few initial successes) is generally poor." With this as underlying thrust, Kreisel proceeds to elaborate on Wittgenstein's "family resemblances of concepts" and "principal pedagogic aim for philosophy", and discusses, in an extended appendix, "proofs and rules" to draw in recent logical experience. With this, Kreisel hovers closest to Wittgenstein's major work, *Philosophical Investigations*.

Kreisel starts by laying some groundwork about (p.15) "General features of traditional philosophy, and some of their implications": (a) "[...] traditional notions occur to us when we know very little." "[...] when we know very little, we tend to see superficial, abstract features of objects. And when we do see specific features we often cannot say very well—cannot 'define' in familiar terms—what we see." (b) "When we know very little, the main intellectual tools available are a sense of coherence and, more generally, introspection." (c)

---

stressed the choice of explicit definitions.

[22]Cf. end of §1.

[23]This recalls Wittgenstein's attempt to compare proofs according to "what we can do with them" in RFM, as already discussed by Kreisel in his RFM review (cf. §1).

"When we know very little compared to the scope of a question, we are often bad at guessing even remotely the methods needed for a satisfactory answer though often we recognize such an answer immediately when we see one." Kreisel peppers (b) and (c) with historical examples involving Galileo, Plato, Aristotle, Newton, and Cauchy.

Kreisel then focuses on (p.17) "Family resemblances of concepts", and, in connection, "the discovery of definitions". It will be remembered[24] that in 1942 Wittgenstein lent young Kreisel a copy of *The Blue Book*, and that, remarkably, Kreisel only reacted about family resemblances, and that with reference to the group concept. Here too, of the many themes of *Philosophical Investigations*, Kreisel concentrates on family resemblances, now with a remarkably literal twist (p.19): "As I see it (now), Wittgenstein's slogan of 'family resemblances' reminds one of a class of phenomena where the limitations of the traditional style are exceptionally vivid, and hence instructive. I mean the phenomena of *literal* family resemblances, say of the Hapsburgs [sic] or the Bourbons [sic]. What can we realistically expect from any definition of such a family resemblance, say, in the style of analytical philosophy?" This focus on literal family resemblances amounts to a local *chiasmus* moving in reverse to Wittgenstein's conceptualization of aspectual similarities and analogues. The tenor of *Philosophical Investigations* is to pursue aspects and work against definitions of family resemblances in terms of biological causes or necessary and sufficient conditions. Be that as it may, Kreisel proceeds to several points (p.19):

(a) "The first thing to expect is, probably, a *genuine theory* of literal family resemblances or some kind of practical mastery. As appears almost certain now, molecular biology is the appropriate tool here." With this scientific coordination of a question emerging "when we know very little", and especially with Kreisel soon following up with "We cannot expect to find a common element in ordinary experience", one can see Kreisel as proceeding orthogonally to Wittgenstein by looking for a genetic reduction. (b) "A second use to expect from a definition would be for the study of our actual *process of recognizing a family resemblance.* At least here, Kreisel is sensing the importance of what Wittgenstein wrote of as "seeing as" and "the dawning of an aspect". (c) (p.21) "An imaginative (clever) definition in this style, in terms of familiar things, may well be useful for *stimulating*—not the actual process of recognition of family of resemblances, but some of its useful results."

---

[24]Cf. §1.

Again taking a scientific approach, Kreisel mentions as an example of this kind of stimulation *"logical validity* in terms of *derivability*, say by Frege's rules".

Lastly, Kreisel attends to what he terms (p.22) "intimate pedagogy", what he took to be (p.14) "Wittgenstein's principal pedagogic aim for philosophy". Kreisel takes a particular tack (p.22): "Suppose we have come to the conclusion that some given notion, for example, one of those grand traditional notions, has to do with a family resemblance [...]. Of course, we do not assume that such a conclusion, even if sound, can be conveyed convincingly, especially to individuals with very limited experience. We ask the pedagogic question: What can be done?" Emphasizing the need for discretion—not to make "grand" claims—and that "precise formalization" can be instructive, Kreisel proceeds to two examples, Tarski's truth definitions and Gödel's incompleteness theorem. Of the latter (p.24) "It unquestionably refutes the idea that, in mathematics, abstract notions are merely used as a *façon de parler*. Hilbert expressed this idea explicitly and precisely in his consistency programme. A more direct formulation of the idea, which is equally easy to make precise, is that a proof by use of abstract notions of a theorem stated in elementary form, can be straightforwardly converted into an elementary proof." Incidentally, Kreisel soon wrote revealingly (p.25): "*Digression* for readers who have seen my (constipated and fumbling) review in [Kreisel, 1958b] of *Remarks on the Foundations of Mathematics*. To me the single most disturbing (and most surprising) defect of those *Remarks* was and remains Wittgenstein's own fumbling."

As in his other articles concerning Wittgenstein, Kreisel insists on taking a scientific approach, and here, in an extended appendix, he further focuses on logic and mathematics to draw subtle distinctions about proofs and rules that round out his contentions. In particular, he has (p.27) "the novel twist of using notions from Brouwer's intuitionist foundations to examine a natural analogue of Church's thesis." While having taken on the motto from the *Philosophical Investigations*, Kreisel interestingly and chiasmatically proceeds away from its broad concerns towards the nuances of progress in logic and mathematics.

[Kreisel, 1983a], "Einige Erläuterungen zu Wittgensteins Kummer mit Hilbert und Gödel", starts out "I was very astonished by the *Remarks on the Foundations of Mathematics* when they came out, especially by those on Gödel's incompleteness theorems, for reasons that I can state precisely only now [...]". The article, in a recapitulative way, engages Wittgenstein's views on consistency and incompleteness with a palatably seasoned appreciation.

Initially, Kreisel adopts and adapts Wittgenstein's (p.296) "proofs easy to take in and remember". In RFM, Wittgenstein had importantly discussed how mathematical proofs are to be "easy to take in and remember (*überschaubar und einprägsam*)" and "perspicuous (*übersichtlich*)". Kreisel declares that "[...] one of the main concerns of mathematics is to provide general guidelines for proofs to be easy to take in and remember." The guidelines are for what he analyzes into two parts as follows: "For usually one starts from a long, opaque proof and dissects it—with intuition—into a few lemmas, that is to say into a structure *easy to take in*. In this process one tries to formulate (or, if necessary to reformulate) the lemmas in such a way that the properties used in their proofs are easily assimilated by the memory, so that they are *easy to remember*." Kreisel frames this with elements from [Kreisel, 1976a] (discussed above), especially the appeal to properties that occur frequently and their axiomatic analysis for perspicuity. He then sets out (p.297): "Now we are ready to apply some of Wittgenstein's favorite slogans to the axiomatic analysis of proofs, e.g. the relatively original: the proof constructs (i.e., in the proof one discovers) new concepts, or the very popular one around 1930: only the proof gives meaning to the theorem that it proves." (It will be remembered that Kreisel in his [1958b, pp.140f] review of RFM had worked chiasmatically against these slogans.) Kreisel proceeds to give "two (entirely elementary) examples",[25] these evidently in the spirit of the elementary RFM examples.

Proceeding to Hilbert's program and consistency, Kreisel declaims (p.298): "Like many others around 1930, Wittgenstein was decidedly enthusiastic about the main component of Hilbert's program: formalization." Yet on two points Wittgenstein was critical: "Firstly, [...] he thought it not fruitful to consider *all* calculations of a 'calculus'. Put differently: formal provability (even by limited means) without regard to ease to take in and remember seemed to him a bad idealization." "Secondly, he was disturbed by Hilbert's exaggerated claims for the importance of consistency." On this last, Kreisel puts Wittgenstein in company with Brouwer and Russell as also "very critical", and mentions Gödel and Gentzen's criticism that "consistency at best guarantees the validity of universal theorems [...], whereas in practice one is rather more interested in existence theorems."

Considering next the shift from provability to proofs, Kreisel writes (p.300): "Since completeness and incompleteness only relate to provability, and have nothing to do with the structure of proofs, they lose their central role." But then, "What happens to incompleteness *proofs* when incompleteness itself loses its 'fundamental' significance? A normal person remembers the good advice: we have nothing to fear but fear itself. In other words, such

---

[25]The first was given in footnote 16.

proofs have more meaningful consequences [...]". On this Kreisel relates an anecdote from the 1940s, the last of his quoted in §1, with how, with the incompleteness proofs, "In Wittgenstein's opinion, Gödel had discovered an absolutely new method of proof."

Kreisel ends with "Wittgenstein's expectations" (pp.301f):

> Above all the *Remarks* were meant to stimulate the reader to have his own thoughts; especially those readers who had already come close to Wittgenstein's thoughts. [...]
>
> This expectation was confirmed by my own experience. When they came out, the *Remarks* did not help me at all. Since the end of the sixties I myself had started to consider structural properties of proofs. After a lecture in 1973 in which I presented these ideas and their development (also by Statman), Nagel drew my attention to the fact that these tendencies (certainly not the details) reminded him of Wittgenstein's *Remarks*. I was absolutely unaware of this connection before then. But I am entirely aware of the additional confidence in my own thoughts that I derived afterwards from leafing through, e.g., Wittgenstein's *Zettel*. Added to this was a certain pleasure, at his skillful formulations and my reformulations of his less skillful ones.

That said, Kreisel retrenches with softer versions of criticisms from his RFM review: how Wittgenstein's specific examples were not fruitful for Kreisel; how he has no use for Wittgenstein's "fussing with *clarity* and *clarification*"; and Wittgenstein's "often erroneous contraposition of *clarification of existing knowledge* and *new constructions*". Nonetheless, throughout [Kreisel, 1983a] there is steady, serious engagement with Wittgensteinian incentives in RFM.

Kreisel's last articles concerning Wittgenstein are variously elliptical, reminiscent, or outright expressionistic. [Kreisel, 1983b] is a quick review of Kripke's 1982 book, *Wittgenstein: On Rules and Private Language*, a review that amounts to a series of chiasmatic remarks putting things in a series of different nutshells. [Kreisel, 1989a] is a collection of "recollections and thoughts" about conversations with Wittgenstein, from which we have already drawn in §1. And [Kreisel, 1989b], *Zu Wittgensteins Sensibilität*, written for a Festschrift, is a remarkably expressionistic series of wide-ranging aphorisms, quips, repartees, and things that came to mind—but nevertheless an article that fully affirms Kreisel's deep engagement with Wittgenstein.

As a way of affirming and accentuating an overall *chiasmus* for Kreisel, his eventual reversal in attitude about RFM after his negative review [Kreisel, 1958b], we consider passages from a appendix to a long letter [1990b] that Kreisel wrote to Grigori Mints in 1990. First, from p.24:

> In a sense I might be said to have made fun of Wittgenstein in a review I wrote in the 50s of his *Remarks on the foundations of mathematics* (although this description certainly does not fit the way I felt about that volume nor about the review). I had made a mistake, which I noticed some 20 years later,* and have referred to it many times. But let me repeat it here, since you may not have taken in these references.
>
> *Main Mistake.* I did not look at the preface, where the editors say in the clearest possible terms that they had found a box full of notes by Wittgenstein, and that *they* had selected what, to *them, seemed most extraordinary.* N.B. I knew those editors! So, if I had looked at the preface this passage would have been an *immediate warning*: what is most extraordinary (= remarkable) to *them* was almost bound to be either wrong [?] or even an aberration.

With the * he references [Kreisel, 1979], a brief review of the second edition [Wittgenstein, 1978] of RFM that does not mention any mistake. In the preface to RFM [Wittgenstein, 1956], the editors nowhere state that they "had found a box full of notes", but do state (p.viii) "[...] what is here published is a *selection* from more extensive manuscripts".

Later on, Kreisel wrote (pp.25f):

> *Consequences of the main mistake.* Actually, in the last paragraph of the review (in the 50s) I said explicitly that I simply did not recognize in the *Remarks on the foundations of mathematics* what I had remembered minimally from my conversations with Wittgenstein. Fittingly (at least from my view of the world), I ignored what I remembered of Wittgenstein, and read the volume as a foil to my then current interests, mentioned above: What, if anything, does it say that is in conflict with—tacitly, the mere coherence of—the foundational tradition?
>
> *STAGGERING OVERSIGHT* on my part. I myself had put on record (in [Kreisel, 1950], somewhere in a highly visible footnote)—published during Wittgenstein's lifetime!—Wittgenstein's perfectly good understanding of Gödel's incompleteness theorem; tacitly, in the mid 40s, after I had explained it to him in $< \frac{1}{2}$ hour in WORDS CONGENIAL TO HIM. In accordance with his habit he recorded the explanation in his own words, incidentally stumbling thereby on—what later came to be called—Henkin's problem.

Henkin's problem is [Henkin, 1952]: "If Σ is any formal system adequate for recursive number theory, a formula (having a certain integer q as its Gödel number) can be constructed which expresses the proposition that the formula with Gödel number q is provable in Σ. Is this formula provable or independent of Σ?" [Kreisel, 1953] discussed an approach to this problem, and then

[Löb, 1955] established provability for more general formulas and under minimal conditions on Σ, the result now known, of course, as Löb's Theorem.[26]

The overall *chiasmus* working its way through the previous and the current sections is *first*, the critical attitude Kreisel took to the *Remarks on the Foundations of Mathematics* in his review [Kreisel, 1958b] as particularly seen in his negative remarks about Wittgenstein's purported construals of the diagonal argument and the first incompleteness theorem, and *second*, a gradual working back, as traced in this section, to a nuanced assessment and appreciation in his articles in the 1970s and 1980s, particularly in [Kreisel, 1983a], ostensibly in tandem with the evolution of his own thinking and experience. The letter cements the *chiasmus* further, working various angles of mistakes and oversights. In particular, Kreisel not only records the *mea culpa* of his not recalling having mentioned Wittgenstein's rule-following version of the diagonal argument in first logic paper [Kreisel, 1950],[27] but credits Wittgenstein for actually formulating Henkin's problem, a problem he himself later worked on.

Stepping back and taking it all in from the beginning, one sees the "early" Kreisel as stirred to his lifelong engagement with constructivity and proof by conversations with Wittgenstein, particularly with the "combinatorial core" of consistency proofs. One sees the "middle" Kreisel with an anxiety of influence reacting negatively in his reviews of Wittgenstein publications, flattening his work on language and insisting on the fruitfulness of research into constructivity and even set theory. Finally, one sees the "later" Kreisel in published essays interestingly integrating his latter-day, seasoned outlook on logic and mathematics with remembrances of the words and ways of Wittgenstein. Proceeding in dialectical engagement, Kreisel growingly acknowledges Wittgenstein as at least providing a conceptual context. But while aspiring to encompass Wittgenstein's broad ways of thinking about language, Kreisel would ultimately remain within the compass of logic and mathematics as set out by RFM. There, the engagement was enlivened by an appreciation of mathematical practice as the place to look for the important structural properties of proof; the role of explicit definitions in that regard; and the importance of proofs "easy to take in and remember". This rounds an arrow of influence, those early conversations with Wittgenstein having stimulated Kreisel to constructivity, logic, and proof.

---

[26]With *Bew* the provability predicate and $\ulcorner \varphi \urcorner$ the Gödel number of φ, Löb's Theorem asserts that for adequate Σ, if $\Sigma \vdash Bew(\ulcorner \varphi \urcorner) \longrightarrow \varphi$, then $\Sigma \vdash \varphi$. Henkin had asked whether $\Sigma \vdash \varphi$ or not in the special, fixed-point case when *is* φ.

[27]Cf. end of §1.

References

[Bolzano, 1817] Bernard Bolzano. *Rein analytischer Beweis des Lehrsatzes, daß zwischen je zwey Werthen, die ein entgegengesetztes Resultat gewähren, wenigstens eine reelle Wurzel der Gleichung liege.* Gottliebe Hasse, Prague, 1817. Translation by Steve Russ in [Ewald, 1996], pages 225–248.

[Cauchy, 1821] Augustin-Louis Cauchy. *Cours d'analyse de l'École royale polytechnique.* Imprimérie royale, Paris, 1821.

[Dedekind, 1872] Richard Dedekind. *Stetigkeit und irrationale Zahlen.* F. Vieweg, Braunschweig, 1872. Translated in [Ewald, 1996], pages 765–779.

[Ewald, 1996] William Ewald. *From Kant to Hilbert.* Clarendon Press, Oxford, 1996.

[Floyd and Mühlhölzer, 2019] Juliet Floyd and Felix Mühlhölzer. Wittgenstein's Annotations to Hardy's *A Course in Pure Mathematics*, 2019. To appear.

[Floyd, 2001] Juliet Floyd. Prose versus proof: Wittgenstein on Gödel, Tarski and truth. *Philosophia Mathematica*, 9:280–307, 2001.

[Floyd, 2012] Juliet Floyd. Wittgenstein's diagonal argument: A variation on Cantor and Turing. In *Epistemology versus Ontology*, pages 24–55. Springer, Dordrecht, 2012.

[Henkin, 1952] Leon Henkin. A problem concerning provability. *The Journal of Symbolic Logic*, 17:160, 1952.

[Kreisel, 1950] Georg Kreisel. Note on arithmetic models for consistent formulae of the predicate calculus. *Fundamenta Mathematicae*, 37:265–285, 1950.

[Kreisel, 1951] Georg Kreisel. On the interpretation of non-finitist proofs—part I. *The Journal of Symbolic Logic*, 16:265–285, 1951.

[Kreisel, 1952a] Georg Kreisel. On the interpretation of non-finitist proofs. Part II. Interpretations of number theory. Applications. *The Journal of Symbolic Logic*, 17:43–58, 1952.

[Kreisel, 1952b] Georg Kreisel. Some elementary inequalities. *Indagationes Mathematicae*, 14:334–338, 1952.

[Kreisel, 1953] Georg Kreisel. On a problem of Henkin's. *Proceedings of the Koninklijke Nederlandse Akademie van Wetenschappen, series A*, 56:405–406, 1953.

[Kreisel, 1958a] Georg Kreisel. Mathematical significance of consistency proofs. *The Journal of Symbolic Logic*, 23:255–182, 1958.

[Kreisel, 1958b] Georg Kreisel. Wittgenstein's *Remarks on the Foundations of Mathematics*. *British Journal for the Philosophy of Science*, 9:135–158, 1958.

[Kreisel, 1960] Georg Kreisel. Wittgenstein's theory and practice of philosophy. *British Journal for the Philosophy of Science*, 11:238–251, 1960. Mainly a review of *The Blue and Brown Books*.

[Kreisel, 1976a] Georg Kreisel. Der unheilvolle Einbruch der Logik in die Mathematik. *Acta Philosophica Fennica*, 28:166–187, 1976.

[Kreisel, 1976b] Georg Kreisel. What have we learnt from Hilbert's second problem? In Felix E. Browder, editor, *Mathematical Developments arising from Hilbert problems*, volume 28 of *Proceedings of Symposia in Pure Mathematics*, pages 93–130, Providence, 1976. American Mathematical Society.

[Kreisel, 1977] Georg Kreisel. On the kind of data needed for a theory of proofs. In Robin O. Gandy and J. Martin E. Hyland, editors, *Logic Colloquium 76*, volume 87 of *Studies in Logic and the Foundations of Mathematics*, pages 111–128. North-Holland, Amsterdam, 1977.

[Kreisel, 1978a] Georg Kreisel. The motto of 'Philosophical Investigations' and the philosophy of proofs and rules. *Grazer Philosophische Studien*, 6:13–38, 1978.

[Kreisel, 1978b] Georg Kreisel. *Wittgenstein's Lectures on the Foundations of Mathematics, Cambridge 1939*. Bulletin of the American Mathematical Society, 84:79–90, 1978. Reprinted in Stuart Shanker, editor, *Ludwig Wittgenstein: Critical Assessments*, 1986, Croom-Helm, London.
[Kreisel, 1978c] Georg Kreisel. Zu Wittgensteins Gesprächen und Vorlesungen über die Grundlagen der Mathematik. In E. Leinfellner, H. Berghel, and A. Hübner, editors, *Proceedings of the Second International Wittgenstein Symposium*, pages 79–81, Vienna, 1978.
[Kreisel, 1979] Georg Kreisel. *Remarks on the Foundations of Mathematics*. American Scientist, 67:619, 1979. Review of the second, 1978 edition.
[Kreisel, 1982] Georg Kreisel. Finiteness theorems in arithmetic: An application of Herbrand's theorem for Σ_2 formulas. In Jacques Stern, editor, *Proceedings of the Herbrand Symposium*, pages 39–55. North-Holland, Amsterdam, 1982.
[Kreisel, 1983a] Georg Kreisel. Einige Erläuterungen zu Wittgensteins Kummer mit Hilbert und Gödel. In *Epistemology and Philosophy of Science. Proceedings of the 7th International Wittgenstein Symposium*, pages 295–303, Vienna, 1983. Hölder-Pichler-Tempsky Verlag.
[Kreisel, 1983b] Georg Kreisel. Saul Kripke, *Wittgenstein on Rules and Private Language*. Canadian Philosophical Reviews, 3:287–289, 1983.
[Kreisel, 1987] Georg Kreisel. Proof theory: some personal recollections, 1987. In [Takeuti, 1987], pages 395–405.
[Kreisel, 1989a] Georg Kreisel. Zu Einigen Gesprächen mit Wittgenstein: Erinnerungen und Gedanken. In *Wittgenstein: Biographie-Philosophie-Praxis, Catalogue for the Exposition at the Wiener Secession, 13 September-29 October 1989*, pages 131–143, Vienna, 1989.
[Kreisel, 1989b] Georg Kreisel. Zu Wittgensteins Sensibilität. In W. Gombocz, H. Rutte, and W. Sauer, editors, *Traditionen und Perspektiven der analytischen Philosophie, Festschrift für Rudolf Haller*, pages 203–223, Vienna, 1989. Hölder-Pichler-Tempsky Verlag.
[Kreisel, 1990a] Georg Kreisel. *About Logic and Logicians: A Palimpsest of Essays by Georg Kreisel*, 1990. A two-volume unpublished collection of essays by Kreisel, selected and arranged by Piergiorgio Odifreddi.
[Kreisel, 1990b] Georg Kreisel. Appendix to a letter to Grigori Mints, 1990. Kreisel made the letter available e.g. to Mathieu Marion.
[Löb, 1955] Michael H. Löb. Solution of a problem of Leon Henkin. *The Journal of Symbolic Logic*, 20:115–118, 1955.
[Mehlman, 2010] Jeffrey Mehlman. *Adventures in the French Trade: Fragments Toward a Life*. Stanford University Press, Redwood City, 2010.
[Monk, 1990] Ray Monk. *Ludwig Wittgenstein: The Duty of Genius*. The Free Press, Macmillan, New York, 1990.
[Odifreddi, 1996] Piergiorgio Odifreddi, editor. *Kreiseliana. About and Around Georg Kreisel*. AK Peters, Wellesley, 1996.
[Putnam, 1975] Hilary Putnam. What is mathematical truth? In Hilary Putnam, *Mathematics, Matter and Method. Philosophical Papers, Volume I*, pages 60–78. Cambridge University Press, Cambridge, 1975.
[Takeuti, 1987] Gaisi Takeuti. *Proof Theory*. North-Holland, Amsterdam, 1987.
[Wittgenstein, 1953] Ludwig Wittgenstein. *Philosophical Investigations*. Basil Blackwell, Oxford, 1953. Edited by G.E.M. Anscombe and R. Rhees and translated by G.E.M. Anscombe.
[Wittgenstein, 1956] Ludwig Wittgenstein. *Remarks on the Foundations of Mathematics*. Basil Blackwell, Oxford, 1956. Edited by G.H. von Wright, R. Rhees, G.E.M. Anscombe and translated by G.E.M. Anscombe. First edition.

[Wittgenstein, 1958] Ludwig Wittgenstein. *The Blue and Brown Books: Preliminary Studies for the 'Philosophical Investigations'*. Basil Blackwell, 1958. Edited by Rush Rhees.

[Wittgenstein, 1976] Ludwig Wittgenstein. *Wittgenstein's Lectures on the Foundations of Mathematics*. Cornell University Press, Ithaca, 1976. Edited by Cora Diamond from the notes of R.G. Bosanquet, Norman Malcolm, Rush Rhees, and Yorick Smythies. Second edition, University of Chicago, Chicago 1989.

[Wittgenstein, 1978] Ludwig Wittgenstein. *Remarks on the Foundations of Mathematics*. Basil Blackwell, Oxford, 1978. Edited by G.H. von Wright, R. Rhees, G.E.M. Anscombe, translated by G.E.M. Anscombe. Second edition with additions.

[Wittgenstein, 1979] Ludwig Wittgenstein. *Wittgenstein's Lectures, Cambridge, 1932–35: From the Notes of Alice Ambrose and Margaret Macdonald*. University of Chicago Press, Chicago, 1979.

[Wittgenstein, 1980] Ludwig Wittgenstein. *Remarks on the Philosophy of Psychology, vol. I*. Chicago University Press, Chicago, 1980. Edited by G.E.M. Anscombe and G.H. von Wright.

[Wittgenstein, 1993] Ludwig Wittgenstein. *Philosophical Occasions, 1912–1915*. Hackett Publishing Company, Indianapolis, 1993. Edited by James C. Klagge and Alfred Nordmann.

2
Kreisel's Dictum

GÖRAN SUNDHOLM

Philosophers, as opposed to mathematicians or mathematical logicians, when considering Kreisel's oeuvre, do not usually consider his technical work in proof theory and constructivism, but have one of three things in mind. The first is his negative review of Wittgenstein's *Remarks on the Foundations of Mathematics*.[1] Secondly, there is what has become known as his "famous squeeze argument" from the London 1965 lecture.[2] Thirdly, and finally, there is, above all, his eponymous *Dictum*. However, even the exact *formulation* of "Kreisel's Dictum" is a matter of considerable intricacy. What does it actually say, and what is its point? Indeed, even concerning its subject matter there are different opinions. It is the purpose of the present note to collect and order the available material, as well as to put on record some novel autobiographical evidence.

The baptism of the dictum took place in 1973, with Michael Dummett officiating at the font:

> (I) Kreisel's dictum, 'The **point** is not the existence of mathematical objects, but the objectivity of **mathematical truth**'.[3]

Since Dummett also gave many alternative formulations of the Dictum I use boldface to indicate identifying characteristics.

---

[1] Kreisel, G. (1958). Wittgenstein's remarks on the foundations of mathematics. *British Journal for the Philosophy of Science* 9 (34):135–158.

[2] Kreisel, Georg (1967). Informal Rigour and Completeness Proofs. In Imre Lakatos (ed.), *Problems in the Philosophy of Mathematics*. North-Holland. pp.138–157. The argument was given prominence—became "famous"—only with an wealth of citations from 2010 onwards, after Hartry Field discussed it in his *Saving Truth from Paradox*, Oxford University Press, 2008, pp.46–48. In pre-internet days, say around 1980, when, I suppose, I was, owing to my dissertation work, as well versed in Kreisel's writings, and their reception, as anyone, the squeeze argument was hardly known and could not be described as famous.

[3] Dummett, Michael, (1975) 'The Philosophical Basis of Intuitionistic Logic', in: H.E. Rose and J.C. Shepherdson (eds.), *Logic Colloquium '73, Proceedings of the Logic Colloquium Bristol July 1973*, North-Holland, Amsterdam, 1975, pp.5–40, at p.19. Reprinted in *Truth and Other Enigmas*, Duckworth, London, 1978, at p.228.

(II) as Kreisel remarked in a review of Wittgenstein, 'the **problem** is not the existence of mathematical objects, but the objectivity of **mathematical statements**'.[4]

(III) as Kreisel has remarked, the **issue concerning platonism** relates, not to the existence of mathematical objects, but to the objectivity of **mathematical statements**.[5]

(IV) As Kreisel remarked apropos of Wittgenstein, the **question** is not whether there are mathematical objects, but whether mathematical **statements** are objective.[6]

(V) As Kreisel has remarked, what is **important** is not the existence of mathematical objects, but the objectivity of mathematical **statements**.[7]

Some observations on these formulations:

1. 1973 is the year of *baptism,* when the dictum becomes *Kreisel's Dictum.* Dummett's Bristol lecture was delivered that year and the Frege book appeared.

2. Dummett's renderings vary:
   the **point**,
   the **issue** concerning Platonism,
   the **problem**,
   the **question**,
   what is important
   and
   the objectivity of mathematical **truth**,
   the objectivity of mathematical **statements**.

3. In (I) and (II) the Dictum is presented as if it were *quoted* from Kreisel.

However, point 3 presents us with a conundrum: *where* in Kreisel's many writings on the Foundations of Mathematics is the *Dictum* found, if at all?

In 1978, in view of the three references (I)–(III) to the *Dictum* in the then just published *Truth and Other Enigmas,* I looked up Kreisel's review of Wittgenstein's *Remarks on the Foundations of Mathematics,*[8] but was unable to

---
[4] *Truth and Other Enigmas*, Preface, p. xxviii.
[5] 'Realism', in: *Truth and Other Enigmas*, a lecture given to an Oxford Society in 1963, and published only in 1978, pp.145–165, at p.146.
[6] *Frege. Philosophy of Language*, Duckworth, London, 1973, Preface, pp. xx–xxi.
[7] *Frege. Philosophy of Language*, p.508.
[8] The reference can be found in the immediately succeeding quotation from myself.

find *"Kreisel's Dictum"* there, and in 1993, in my talk at the First Mussomeli Conference on the Philosophy of Michael Dummett, where Dummett was due to give written reply to our papers, I even ventured into print regarding my scepticism on the matter:

> Prof. Dummett has given wide currency to Kreisel's dictum: "The point is not the existence of mathematical objects, but the objectivity of mathematical statements" [...], concerning which questions were set in the B. Phil examinations during my time at Oxford. Prof. Dummett refers to a review of Wittgenstein for the quote in question. What Kreisel states in footnote 1 on p.138 of his review of Wittgenstein's 'Remarks on the Foundations of Mathematics', in the British Journal for the Philosophy of Science, Vol. no. 9, 1958, is that "Wittgenstein argues against a notion of a mathematical object (presumably: substance), but, at least in places [...] not against the objectivity of mathematics, especially through the recognition of formal facts". *Se non è vero, è ben trovato* [...][9]

Dummett's "Comments on Sundholm", alas, were concerned solely with my fifth Vestige of Realism and did not respond to the Dictum issue. Sadly I never took the opportunity to ask him about it later. Be that as it may, *"Kreisel's Dictum"* has received fairly wide currency over the years and many have searched for a location in Kreisel's work.[10]

In 2001 and 2011 there were lengthy exchanges on the FOM list with many participants, among whom Tait, Urquhart, Reck, Hazen, Ratikananen, Palma, Franchella...[11]

In one of these exchanges, on the FOM list, May 11, 2001, while writing "[f]rom a warm and fragrant South Bend", Mic Detlefsen, gave helpful hints

> In the 1958 review that Bill [Tait] mentioned, there is a remark that relates to this thread. On p.138, note 1, Kreisel writes:
>
> "[...] it should be noted that Wittgenstein argues against a notion of

---

[9] Göran Sundholm, 'Vestiges of Realism', in: Brian McGuinness and G. Oliveri, *The Philosophy of Michael Dummett*, Kluwer, Dordrecht, 1994, pp.137–165, p.144, footnote 24.

[10] Thus, for example, Crispin Wright in his *Wittgenstein on the Foundations of Mathematics*, Duckworth, London, 1978, p.5, footnote 1, gives "54" as a reference for Kreisel's Dictum (though not under that name). That reference, without a further page indication, is to Kreisel's long survey paper 'Mathematical Logic' in T.L. Saaty, *Lectures on Modern Mathematics, Vol.III*, John Wiley and Sons, New York, 1965, pp.95–195. In spite of many careful readings I have not been able to find there anything like the Dummettian formulation that Wright deploys.

[11] https://cs.nyu.edu/mailman/listinfo/fom/

mathematical object (presumably: substance), but, at least in places [...] not against the objectivity of mathematics, especially through his recognition of formal facts [...]"

He makes similar or related remarks in the following places:

(i) p.97 of his 1965 paper "Two notes on the foundations of set theory".

(ii) pp.219f of his 1965 paper "Mathematical logic: what has it done for the philosophy of mathematics?"

(iii) p.20 of his 1970 paper, "The formalist-positivist doctrine of mathematical precision in the light of experience".[12]

However, also in these locations one is not able to find the *Dictum* as formulated by Dummett.

To confuse matters a bit more, there are also other pithy formulations of Kreisel's, on the interplay between classical and intuitionistic mathematics, that, on and off, have been given the status of a *dictum*. For example, Peter Pagin considered another Kreiselian apothegm in the Philosophy of Mathematics, namely

> The conception that the proof relation is decidable is captured by Georg Kreisel's dictum 'We can recognise a proof when we see one',[13]

whereas I myself have often quoted what I labelled Kreisel's "*Bucharest* **Question**" from LMPS IV, 1971

> Was the (logical) language of current intuitionistic systems obtained by uncritical transfer from languages which were, tacitly, understood classically?[14]

---

[12]G. Kreisel, 'Two notes on the foundations of set theory', *Dialectica*, Vol.23, No.2 (1969), pp.93–14; G. Kreisel, 'Mathematical Logic: What has It Done for the Philosophy of Mathematics?', in: R. Schoenmann, *Bertrand Russell, Philosopher of the Century*, Allen and Unwin, 1967, pp.201–272; G. Kreisel, The formalist-positivist doctrine of mathematical precision in the light of experience. *L' Age de la Science*, vol.3 (1970), pp.17–45.

Note here that Detlefsen erroneously gives 1965 as the date for the second article.

[13]G. Kreisel 'Foundations of Intuitionistic Logic', *Logic, Methodology and Philosophy of Science. Proceedings of the 1960 International Congress* (E. Nagel, P. Suppes, and A. Tarski, eds.), Stanford University Press, 1962, pp.198–210, at p.209.

Peter Pagin, 'Compositionality, Understanding, and Proofs' *Mind*, 118 (2009), 713–737, at p.720.

[14]G. Kreisel, 'Perspectives in the Philosophy of Pure Mathematics', in: *Proceedings of the Fourth International Congress for Logic, Methodology and Philosophy of Science*, Bucharest, 1971 (eds. P. Suppes, Leon Henkin, Athanase Joja, Gr.C. Moisil), North-Holland, Amsterdam, 1973, pp.255–277, at p.268.

Regarding *where* Kreisel offered his *Dictum*, Thomas Forster, on FOM, May 10, 2001, very sensibly asked a pertinent question that was also an implicit suggestion:

> Has anyone thought of actually asking Kreisel? Or would that be *lèse majesté*?

As a matter of fact someone did think of this, namely **myself**.

In early April 1978, at the suggestion of Dana Scott (who was standing in as my supervisor for the absent Robin Gandy) I went to Paris to offer Kreisel a copy of my B. Phil. dissertation on the omega-rule. The first meeting, at *Institut des hautes études scientifiques*, the mathematical research centre at Burres-sur-Yvette, went very well; Kreisel received me with Austrian courtesy, excellent coffee, and delicate small pastries, and invited me to come back the following day (perhaps also because I could then, as he suggested and requested, collect a tax-form for him at the US consulate next to Place de la Concorde). There was, relatively speaking, a lot of time for scientific conversation, over the two consecutive days, so my dissertation was not the sole topic that was raised. From my reading of Dummett, the difficulty was fresh in my mind of not being able to locate the *Dictum* anywhere in Kreisel's writings. At the time I probably knew them as well as anybody, owing to my dissertation work on the omega rule, and here was the opportunity to find out. So I asked Kreisel about it.

Alas, I did not get a pat quotation from him. On the contrary, Kreisel did **not endorse** any of the formulations as given by Dummett.

1. He did remark, however, in an arresting and very Kreiselian phrase that has vividly stuck in my memory, that linking language and objectivity along the lines of the various Dummettian formulations of the *Dictum*,

    > provides rich possibilities for speculative metaphysics and should delight a professional philosopher like Dummett.

2. Kreisel also explicitly **rejected** my proposed emendation of the *Dictum* into:

    > The point is not the existence of mathematical objects, but the objectivity of mathematical **proofs**.

    This, he said, would lead, straight on to "*operational semantics*" and "we both know what I think about that".

The barbed sting in that final observation is directed against Paul Lorenzen and Dag Prawitz, and refers to the Appendix I of *A Survey of Proof Theory II*.[15] The waspishness returns also in other passages, such as the following that is taken from an article that I knew almost by heart, owing to my dissertation work, and about which Kreisel and I talked a fair bit:

> It is fair to say that one of the most useful immediate consequences of the shift from (i) sequent formulations preferred by Gentzen and Takeuti to (ii) the natural deduction formulations preferred by Prawitz was just this: in contrast to (i), in case (ii) a particular mapping suggested itself; so there remained *a* well-defined normalization problem even after the corresponding normal form theorem had been established by model theoretic methods (which ensured at least one normalization procedure, the trivial ρ above). Five years later we cannot be satisfied with such virtues: *quod decet bovem dedecet Jovem*.[16]

So at the time I reached the conclusion that "*Kreisel's Dictum*" really is **Dummett's**. This gains plausibility when one considers that Dummett seems to have written books and articles largely from memory, without detailed consultation of sources. Thus he reports that the large Frege book was essentially written from memory;

> [W]hile writing the book, I found that it gravely impeded the flow of composition for me to stop to locate a particular remark I was citing from Frege, [...], and, to avoid this, I adopted the policy of writing from memory [...].[17]

Furthermore, as Dummett remarked in an interview with Joachim Schulte:

> One effect of quoting from memory without looking up the passage is that you overlook the context and thereby the connections that the author makes and that ought to be pointed out and discussed.[18]

---

[15] G. Kreisel, 'A Survey of Proof Theory II', in: J.E. Fenstad (ed.) *Proceedings of the Second Scandinavian Logic Symposium*, North-Holland, Amsterdam, 1971, pp.109–170, Appendix I, at pp.149–165.

[16] G. Kreisel, G. Mints, and S. Simpson, 'The Use of Abstract Language in Elementary Metamathematics: Some Pedagogic Examples' in: Rohit Parikh, ed., *Logic Colloquium*, Springer Lecture Notes in Mathematics, Vol.453 (1975), pp.38–131, at p.45.

[17] *Frege. Philosophy of Language* (second edition), Duckworth, London 1981, Preface, p. xiv.

[18] Michael Dummett, *Origins of Analytical Philosophy*, Duckworth, London, 1993, Appendix: Interview, pp.167–195, at p.172.

Kreisel's Dictum

In the same interview, Dummett tell us how he had composed the Review of Wittgenstein's *Remarks*:

> I in fact reviewed the *Remarks on the Foundations of Mathematics* when it first came out [...] I had tried to write this review in the usual way, with the book beside me and looking up passages in it, and I found that I could not do it.
>
> [...]
>
> I found it all crumbling in my fingers. So I put the book away and deliberately thought no more about the review for about three months. Then, with my now impaired memory of the book, I wrote the review—deliberately without opening the book again.
>
> [...]
>
> Finally I inserted some references.

In the light of this method of composition it seems to me quite likely that Kreisel's actual formulation in 1958, with its large burden of inserted page references

> Wittgenstein argues against a notion of a mathematical object (presumably: substance), but, at least in places (p.124, p.35 or p.96, p.71, lines 5 and 4 from below) not against the objectivity of mathematics, especially through the recognition of formal facts (p.128, p.50).

may well, after a decade and a half have been transformed, by Dummett's "by then impaired memory" into either of the pithy forms I–IV of the Dictum that are given above.

Why did Dummett introduce "Kreisel's Dictum"? If we avail ourselves of a "Platonist" ontology, then a rightness-norm that ensures *objectivity in the sense that not everything goes* is readily available in terms of that ontology; how matters stand in the ontology—the obtaining of "states of affairs"—determines what propositions are true (and of course also which are false), and so, against this background, our judgements, our epistemic acts of knowledge, can derail and may be wrong, namely when such an act ascribes truth to a proposition that presents a state of affairs that does not obtain and is in fact false. So here the ontology provides a norm of objectivity: given an objective ontology that imposes bivalence on propositions, via a correspondence theory account of propositional truth, it is possible to go wrong relative it.

The following diagram (that I have used on many occasions) spells out the possibilities.[19]

OBJECTS AND OBJECTIVITY: holding open the possibility for mistakes

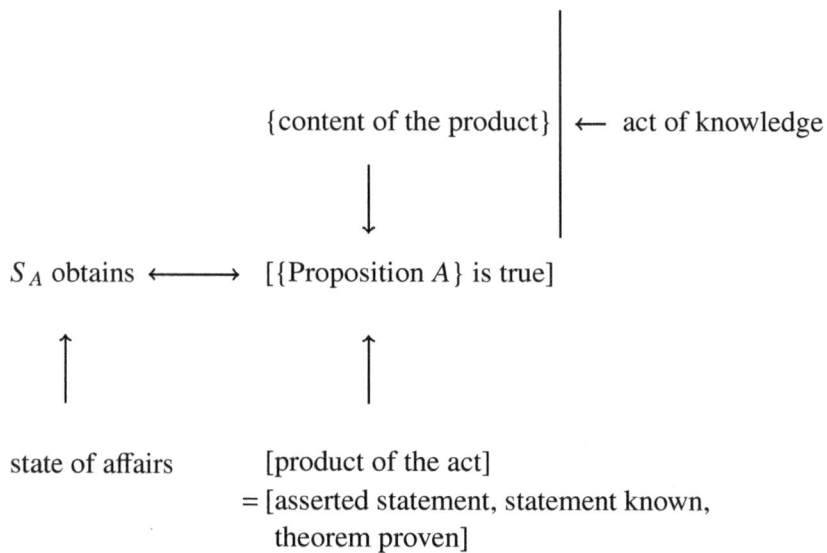

At each level there operates what German philosophers call a suitable notion of *Geltung*, that we may perhaps render as a "correctness notion":

1. The act is *right*;

2. The act product, or object of the act, the judgement made, is *correct*;

3. The content, the proposition, is *true*;

4. The ontological correlate, the state of affairs, *obtains*.

The *sine qua non* minimum requirement on an epistemological position is the notion of Rightness at level (1). Without it anything goes, and it is impossible to be mistaken: nothing is objective anymore. "*Alors, si Dieu est mort tous est permis.*" That is, without a norm, everything is permitted. A Platonist will conveniently start with the ontological notion at (4) and accept bivalence there:

---

[19] See, for instance, Göran Sundholm, 'Antirealism and the Roles of Truth', in: M. Sintonen, J. Wolenski & I. Niiniluoto (eds.), *Handbook of Epistemology*, Kluwer, Dordrecht, 2004, pp.437–466, at p.444.

a state of affairs either does or does not *obtain* (German *bestehen*). By "exporting Bivalence" in the search for objectivity, the Platonist accordingly obtains that the proposition that presents the state of affairs is true or false, whereas the theorem (the demonstrated assertion) that states that the proposition is true is correct when that proposition really is true. Finally, the act is right, when its product is correct. (This position is an epistemological pendant to Consequentialism in Ethics, where the moral goodness of an act is reduced to certain properties pertaining to the *result* of the act.) Wittgenstein's **Tractarian Realism** is as fine an example of the above diagram as any: the obtaining/non-obtaining with respect to states of affairs (German *Sachverhalte*) is the pivot around which the logic of the *Tractatus* revolves.

However, we are not obliged to follow the Realist in taking the ontology as fundamental to epistemic objectivity. It is here, in this context that "Kreisel's Dictum" fulfills the valuable task of pointing to the open the possibility that the required rightness norm for our acts of knowledge will not be held accountable towards a pre-existing Platonist ontology, but might be given in another way, or even considered *sui generis*. Brouwer (*and* Bishop Berkeley) are what I have called "*Ontological Descriptivists*" in my joint Cerisy paper with Van Atten.[20] Both reduce truth of propositions, correctness of assertions, and rightness of knowledge-acts to matters of ontology, *in casu* an *idealist* ontology of "mental objects". Thus they are ontological *idealists*, but with respect to this *idealist ontology*—that from a Platonist point of view is considered "deviant"—they are *epistemic realists*: how matters stand in that deviant ontology has the last word on questions of truth and knowledge. This, then, is the role of Kreisel's Dictum: to make us aware of the possibility of having objective truth and meaning that is not based on the ontology.

In fact both Kreisel and Dummett are ontological descriptivists. Kreisel's original Theory of Constructions from the Stanford LMPS congress was construed as a metamathematical interpretation, and does not offer meaning explanations as a Dummettian Theory Of Meaning.[21] The Theory of Constructions has a basic relation between constructions and propositions $\Pi(c, A)$, read "construction c is a proof of proposition A", and axioms that regulate how constructions may be built out of other constructions. Since "we recognize

---

[20] Sundholm G., van Atten M. 'The proper explanation of intuitionistic logic: on Brouwer's demonstration of the Bar Theorem'. In: van Atten M., Boldini P., Bourdeau M., Heinzmann G. (eds.) *One Hundred Years of Intuitionism (1907–2007)*. Publications des Archives Henri Poincaré / Publications of the Henri Poincaré Archives. Birkhäuser, Basel, 2008, pp.60–77.

[21] Kreisel, 'Foundations of Intuitionistic Logic', op. cit., footnote 13. Properly speaking the Π-predicate is three-place in Kreisel's theory, but for present purposes my simplified notion may suffice.

a proof when we see one" the relation is decidable in the formal constructive sense: for any proposition A, it is a theorem of the Theory of Constructions that $\forall c[\Pi(c, A) \vee \neg \Pi(c, A)]$. Intuitionistic predicate logic and arithmetic are both formally embedded into this theory, in much the same way that arithmetic is formally embedded into Zermelo set theory.[22] Against the background of the Dictum it is somewhat ironic that this Kreiselian theory constructions appears to be just another ontologically construed account with just another universe of (construction-)objects that replaces the Platonist universe of set theoretic objects.

Also Dummett succumbs to the charms of Objectivist Descriptivism: since equations between number-theoretic terms in the language of arithmetic are formally decidable, in the sense that constructively one readily demonstrates $\forall x : N[A(x) \vee \neg A(x)]$, where $A(x)$ is quantifier free, the constructivist and Platonist both assert the same quantifier-free closed sentences, and so, according to Dummett, they mean they same for both.[23] Here I beg to cavil: for instance the decidable equation, $5 + 2 = 7$ has its meaning given constructively in term of the recursion equations for $+$, and the explicit definitions of 2, 5, and 7 in terms of successor and 0. Platonistically, on the other hand, the $+$ function, that is, the infinite set of ordered pairs

$$M = \{\langle\langle m, n\rangle, m + n\rangle \mid m, n \in N\}$$

is an element of the cumulative hierarchy at an early infinite stage $V(\omega + k)$, for a suitable $k \in N$, and the sentence $5 + 2 = 7$ has the meaning

$$\langle\langle 5, 2\rangle, 7\rangle \in M.$$

Both Kreisel, in his Theory of Constructions, and Dummett, in his view on the meaning of quantifier-free statements seem, seem to have fallen prey to the tendency against which Kresiel warned in his Bucharest Question: the formal systems of constructivism have been too closely modelled on their Platonist counterparts.

---

[22] My 'Constructions, proofs, and the meaning of the logical constants', *Journal of Philosophical Logic*, 12 (1983), pp.151–72, §1, goes into Kreisel's formalization of Heyting's work in some detail. Walter Dean and Hidenori Kurakawa, 'Kreisel's Theory of Constructions, the Kreisel–Goodman Paradox, and the Second Clause', in: T. Piecha and P. Schroeder-Heister (eds.), *Advances in Proof-Theoretic Semantics*, Springer, Dordrecht, 2016, pp.27–63, aim to "consider the prospects for developing a consistent variant of the Theory of Constructions originally proposed by Georg Kreisel".

[23] Dummett, 'Philosophical Basis...', *Truth and other Enigams*, pp.230–231, op. cit., footnote 3.

My purpose here has been only the modest one to point to different readings of "Kreisel's Dictum" and to put the information from my 1978 conversation with Georg Kreisel on the record. The challenging task of providing, in the first instance, mathematical language with an account of meaning and truth that is objective, but is not reducible to the relevant ontology, will have to be left for another occasion.[24]

---

[24] In 'Antirealism and the Roles of Truth', op. cit, footnote 19, §1, and in 'Error', *Topoi*, 31 (2012), pp.87–92, I have given some indications of how the traditional Coherence, Consensus, and Pragmatic Theories of Truth can be seen as instruments for filling in the Rightness norm at the level of acts, in the face of an inconsistency or incoherence. They then serve as tools for deciding, among many candidates, which act(s) of knowledge to retract.

3

Local Formalizations in Nonlinear Analysis and Related Areas and Proof-Theoretic Tameness

ULRICH KOHLENBACH

> The object lesson concerns the passage from the foundational aims for which various branches of modern logic were originally developed to the discovery of areas and problems for which logical methods are effective tools. The main point stressed here is that this passage did not consist of successive refinements, a gradual evolution by adaptation as it were, but required radical changes of direction, to be compared to evolution by migration.
>
> [...] there is plenty of scope for specialist experience in logic provided (i) new questions are asked and (ii) that experience is combined with more specific knowledge. [Kreisel, 1985b, pp.139f]

1 Introduction

In the recent book [2018], John Baldwin stresses that while early 20th century logic focused on the foundation of all of mathematics, 'contemporary model theory makes formalization of *specific mathematical areas* a powerful tool' and uses 'local formalizations for distinct mathematical areas in order to organize and do mathematics, and to analyze mathematical practice'. Moreover, 'geometry [...] plays a fundamental role in analyzing the models of tame theories and solving problems in other areas of mathematics' ([2018, p.3]). As a result of this 'paradigm shift', model-theoretic methods became a useful tool in core areas of mathematics such as algebra or algebraic geometry.

Baldwin ([2018, p.250]) cites [Kreisel, 1985a] in this respect e.g. 'Kreisel had identified one element of the malaise: "a preoccupation with a universal framework (a universal language, for example) and thus with logical possibilities. This preoccupation is at heart of the malaise; it concerns a potential conflict between pursuing these logical ideals and effective knowledge"'. However, where Kreisel is skeptical about whether model-theoretic 'transfer principles must take the literary form of metatheorems' (see [1985a] as quoted in [2018, p.62]), Baldwin ([2018, p.62]) states that 'given the results discussed in this book, we see Kreisel as overly pessimistic about the prospects of metatheorems'.

By its historical origin, proof theory has been particularly focused on general foundational issues such as the consistency strength of whole systems of analysis and set theory which extend basic number theory and so are not 'tame' in the model-theoretic sense. Rather than ruling out the possibility of 'Gödelian phenomena' by considering only tame theories one goal in recent decades has been to actually produce such phenomena in the context of ordinary mathematics. One famous example for this is Harvey Friedman's discovery that a finitary (Π_2^0-form) of Kruskal's theorem cannot be proven in predicative mathematics (in the sense of ATR_0). Many much stronger forms of such 'concrete incompleteness phenomena' have been discovered by Friedman during the last decades.

A different development, again initiated by Harvey Friedman [1976] and then developed mainly by Stephen Simpson [1999] and his collaborators under the name of 'reverse mathematics', has been to investigate the proof-theoretic strength of basic theorems used in core mathematics relative to a weak base system RCA_0 of second-order number theory which is Π_2^0-conservative over primitive recursive arithmetic PRA. One of the outcomes of this line of research has been that most of that part of existing ordinary mathematics (as long as it is formalizable in the language of 2nd order number theory) can be carried out already in systems such as ACA_0 or WKL_0 which are conservative over Peano arithmetic PA resp. the fragment of PA with Σ_1^0-induction (which in turn is Π_2^0-conservative over primitive recursive arithmetic PRA). This already indicates that a substantial amount of ordinary mathematics is tame in a proof-theoretic sense even when in principle non-tame structures such as a natural numbers are present.

While also reverse mathematics deals with *provability* in certain formal systems (though addressing theorems from ordinary mathematics), Kreisel asked already many decades ago for a shift of emphasis towards the 'unwinding' of *specific proofs*. Moreover, here the focus is on proofs of theorems A which have a *simpler logical form* than the noneffective set-theoretic tools used in the proof (and, in particular, are not equivalent to the latter in sense of reverse mathematics). In fact, the combinatorial or numerical nature of the theorems A considered naturally asks for explicit witnesses or bounds to be extracted from the prima facie noneffective proof. To use proof-theoretic methods for this new type of purpose has been proposed by Kreisel since the 50's, most specifically in his 'unwinding of proofs' program. This has been carried out under the name of 'proof mining' since around 2000 systematically in the area of nonlinear analysis such as fixed point theory, ergodic theory, abstract Cauchy problems, nonlinear semigroups, convex optimization

Local Formalizations in Nonlinear Analysis

and geodesic geometry (see [Kohlenbach, forthcoming] for an account of the vast influence Kreisel's insights have had on prompting this development and [Kohlenbach, 2008a, 2017] for surveys on the results obtained in the proof mining paradigm).

Many theorems in nonlinear analysis concern convergence results for iterated procedures (x_n) for the computation of fixed points of some mapping $T : C \to C$ (C typically being a convex subset of some normed or geodesic space), zeros or minimizers of mappings $F : C \to \overline{\mathbb{R}}$ etc.

Here one usually works in the context of abstract classes of normed and metric structures (used as parameters of the problem) which are not assumed to be separable and so cannot even be expressed in the language of 2nd order number theory. In fact, it is the very absence of any separability assumptions which makes it possible to obtain highly uniform bounds which only depend on general local metric bounds without any compactness assumptions required.

For so-called asymptotic regularity results $d(Tx_n, x_n) \to 0$ one typically obtains full rates of convergence which are mostly polynomial in the relevant data and moduli of the problem (see e.g. [Kohlenbach, 2019] for a polynomial rate of convergence which has been extracted from Bauschke's [2003] solution of the 'zero displacement conjecture') or at least are simple exponential (as e.g. in [Kohlenbach et al., 2017], which analyzes an asymptotic regularity proof from [Ariza-Ruiz et al., 2015] that uses iterated arithmetical comprehension, and [Nicolae, 2013]; see also below). For strong convergence theorems for (x_n) itself one usually can show that (unless for special cases where e.g. the uniqueness of the fixed point of T can be established or one has a so-called modulus of regularity) there is no computable rate of convergence even for simple cases such as $C := [0, 1]$ and easily computable mappings T (see e.g. [Neumann, 2015]). So one usually only has effective so-called rates $\Phi(\varepsilon, \underline{a}, g)$ of metastability (in the sense of Tao [2008a, 2008b])

$$\forall \varepsilon \in \mathbb{Q}_+^* \, \forall g : \mathbb{N} \to \mathbb{N} \, \exists N \leq \Phi(\varepsilon, \underline{a}, g) \, \forall i, j \in [N, N + g(N)] \, (d(x_i, x_j) < \varepsilon)$$

which is an instance of Kreisel's no-counterexample interpretation applied to the Cauchy property (see [Kreisel, 1951, 1952]). These are—with very few exceptions—of the form

$$(+) \; \Phi(\varepsilon, \underline{a}, g) = \left(\chi_1(\varepsilon, \underline{a}) \circ \tilde{g} \circ \chi_2(\varepsilon, \underline{a}) \right)^{(B(\varepsilon, \underline{a}))}(0),$$

where χ_1, χ_2, B are simple (typically polynomial) functions in ε and (majorants of) the parameters $\underline{a}$ of the problem involved but which do not depend on g (here $\tilde{g}(n) := \max\{g(i) : i \leq n\} + n$) and $f^{(n)}(0)$ denotes the n-th iteration of f.

The need for such an iteration usually is due to the arithmetical residuum of a use of sequential compactness which even in the case of a single use of the convergence of bounded monotone sequence of reals uses Σ_1^0-induction Σ_1^0-IA which suffices to show the totality of function iteration (see also below). In fact, the Cauchy property of monotone sequences in [0, 1] is equivalent to Σ_1^0-IA (see [Kohlenbach, 2000]).

So despite of the fact that convergence theorems of the form above crucially use quantification over natural numbers and so per se could display phenomena of enormous growth rates (as in Kruskal's theorem) it is an *empirical* fact, though, that with a few notable exceptions, proofs e.g. in analysis seem to be tame in the sense of allowing for the extraction of bounds of rather low complexity. It is this frequent *proof-theoretic tameness* in currently existing ordinary mathematics which makes the program of unwinding proofs so rewarding but which to diagnose requires a proof-theoretic analysis in each particular case.

Very recently, we analyzed a proof for a central strong convergence result in nonlinear analysis where—as it stands—the rate of metastability for the first time uses primitive recursion of type 1, i.e. the fragment T_1 of Gödel's T whose definable type-1 functions coincide with the provably total functions of the fragment of Peano arithmetic with Σ_2^0-induction (which is equivalent to Π_2^0-induction) and e.g. includes the Ackermann function. Only future research will show whether this is an artefact of the proof being analyzed (or whether even a closer examination of the extracted bound allows for a T_0-definition) or is best possible.

In each of the applications of proof mining it is the bound extracted and/or the general mathematical insights (also qualitative ones such as generalizations to geodesic settings and abstract versions, see e.g. [Leuştean et al., 2018]) into the mathematical situation at hand which is of interest rather than to add to the 'security' of the original proof: 'Of course, being special, finitist proofs do have some special properties including virtues. It just so happens that (special) reliability is not among them.' ([Kreisel, 1985b, p.145]). One of these virtues e.g. often is the fact that the analyzed proofs suggests immediate generalizations (e.g. to geodesic setting or more general classes of geodesic spaces etc., see e.g. [Kohlenbach and Leuştean, 2003] and [Kohlenbach et al., 2017] or [Leuştean and Nicolae, 2016] which generalizes the analysis from [Kohlenbach and Leuştean, 2012] from CAT(0) to CAT(κ)-spaces for $\kappa > 0$).

2 General observations made in case studies I: extensionality

As mentioned already (and discussed in detail e.g. in [Kohlenbach, 2005a, 2008b] and [Gerhardy and Kohlenbach, 2008]) most situations in nonlinear analysis involve abstract classes of normed or metric structures such as Hilbert spaces, uniformly convex spaces or CAT(κ) spaces which are determined by general geometric conditions while not assuming separability. The significance of the latter does not primarily rest on the greater generality but on the fact that if a proof does *not use separability* of the metric structures X the extracted *bounds are highly uniform*, i.e. independent from norm-bounded or metrically bounded parameters from X without any compactness assumption. This uniformity is of interest also in cases where the extracted bound is applied only to separable or even (boundedly) compact structures, e.g. by providing bounds which are independent of the dimension of the space under consideration.

In order to get the appropriate input data for the bounds, one may have to enrich X by suitable moduli e.g. of convexity or smoothness etc. Specially designed *logical metatheorems* which are *tailored* for the class of statements at hand guarantee for whole classes of proofs the extractability of uniform bounds which only depend on X via these moduli (compare Baldwin on 'local' versus 'global' formalizations in contemporary model theory). For the formalization of proofs in the context of abstract metric structures X we use systems formulated in the language of all finite types ρ over the base types $\mathbb{N}$ and X.

The structures admissible in these metatheorems must be axiomatizable by axioms which have a simple monotone functional interpretation (in the respective moduli). Usually, this is guaranteed by verifying that, given maybe some modulus $\omega : \mathbb{N} \to \mathbb{N}$, the axioms stating that X belongs to the respective class of structures (with modulus ω) can be expressed in purely universal form. This e.g. is the case for the following classes of spaces: metric, hyperbolic, CAT(0), CAT($\kappa > 0$), normed spaces, their completions, Hilbert, uniformly convex, uniformly smooth (not: separable, strictly convex or smooth) spaces, abstract L^p- and $C(K)$-spaces and all normed structures axiomatizable in positive bounded logic (Henson, Ben-Yaacov etc., see [Günzel and Kohlenbach, 2016]).

Similarly, the conditions on the classes of admissible mappings T need to be axiomatizable in this way (again possibly using suitable moduli $\omega : \mathbb{N} \to \mathbb{N}$) which includes the following classes: uniformly continuous, Lipschitz-continuous, nonexpansive, firmly nonexpansive, strongly (quasi-)nonexpansive, pseudo-contractive mappings, directionally nonexpansive mappings, mappings satisfying Suzuki's condition (E), maximally monotone and accretive etc. mappings where in the latter two cases also set-valued operators $T : X \to 2^X$ have been treated.

Some of these conditions imply the uniform continuity of the operator T which then in turn implies the *extensionality* of T

$$(*) \; \forall x, y \in X \, (x =_X y \to T(x) =_X T(y)),$$

where $x =_X y :\equiv d_X(x, y) =_\mathbb{R} 0$. This extensionality must not be included as an axiom into the formal systems for which a metatheorem on uniform bound extractions can be expected to hold since, otherwise, the very statement of these metatheorems (on the extractability of uniform bounds) would imply the uniform continuity of T together with a modulus of uniform continuity (even largely independent of T). In fact, uniform continuity *is* the uniform quantitative version of extensionality. So when using conditions on T such as pseudo-contractivity, quasi- or directional nonexpansivity or Suzuki's condition (E) one cannot make free use of the extensionality axiom in formalizing a given proof. The principle $(*)$ (and a fortiori its extensions to higher types) appears to be the *only* principle used in mathematics which per se does not have any computational content. How then is this addressed in the practice of proof mining? Here roughly three different situations can occur:

1. The use of extensionality can be seen to be an instance of the (admissible in the aforementioned metatheorems) quantifier-free *rule* of extensionality

$$\frac{A_{qf} \to s =_X t}{A_{qf} \to T(s) =_X T(t)},$$

where A_{qf} is a quantifier-free formula which may contain parameters.

2. There is an essential use of the extensionality axiom but of a special form whose uniform quantitative version does not require a modulus of uniform continuity. A particularly important instance of this is the use of extensionality in the form

$$\forall x, y \in X \, (x =_X y \wedge Tx =_X x \to Ty =_X y)$$

whose uniform quantitative form only requires so-called moduli of uniform closedness $\delta_T, \omega_T : \mathbb{N} \to \mathbb{N}$ such that

$$(*) \begin{cases} \forall x, y \in X \, \forall k \in \mathbb{N} \, \left(d(x, y) < \frac{1}{\omega_T(k)+1} \wedge d(x, Tx) < \frac{1}{\delta_T(k)+1} \right. \\ \left. \to d(y, Ty) \leq \frac{1}{k+1} \right) \end{cases}$$

which e.g. can easily be constructed if T satisfies Suzuki's condition (E) even though the latter does not imply the continuity of T (see [Kohlenbach et al., 2018a] and [Kohlenbach, 2017]).

3. If the extensionality axiom is needed but the properties of T do not imply the existence of a uniform bound on the resp. use of extensionality then such a modulus needs to be added as an assumption (a modulus of uniform continuity always suffices but weaker moduli as in the item above are often sufficient, see Proposition 4.15 and Remark 4.16 in [Kohlenbach, 2016]).

As is evident from the above, the proof-theoretic bound extraction methods do *not* generally require all functions involved to be uniformly continuous which is the common assumption in continuous or positive bounded logic (see, however, the recent paper [Cho, 2016]).

3 General observations made in case studies II: noneffective existence principles

The main noneffective existence principles applied in nonlinear analysis can be divided in the following two classes:

1. Principles which have the form of a so-called axiom

$$\Delta : \forall x^{\delta} \exists y \leq_{\rho} sx \forall z^{\tau} A_{qf}(x, y, z),$$

where s is some closed term of the system used and A_{qf} a quantifier-free formula. Here $u \leq_{\rho} v$ is pointwise defined, where if u has the type $\rho = \rho_1 \to (\ldots \to (\rho_k \to \tau)\ldots)$ with $\tau \in \{\mathbb{N}, X\}$ then v has the type $\rho_1 \to (\ldots \to (\rho_k \to \mathbb{N})\ldots)$ and $u(\underline{w}) \leq_X v(\underline{w}) :\equiv \|u(\underline{w})\| \leq v(\underline{w})$ in the normed case and $u(\underline{w}) \leq_X v(\underline{w}) :\equiv d_X(u(\underline{w}), a) \leq v(\underline{w})$ for some reference point $a \in X$ in the metric case (see [Kohlenbach, 2008b] and [Günzel and Kohlenbach, 2016]).

2. Principles such as (strong as well as weak) sequential compactness and the existence of projections (metric as well as sunny nonexpansive ones) as well as Banach limits which use arithmetical comprehension, sometimes prima facie also in its uniform version

$$(\exists^2) : \exists \varphi \forall f^{\mathbb{N} \to \mathbb{N}} (\varphi(f) =_{\mathbb{N}} 0 \leftrightarrow \exists n^{\mathbb{N}} f(n) =_{\mathbb{N}} 0)$$

(see further below) or arithmetical dependent choice (and in the case of Banach limits even the existence of nontrivial ultrafilters).

If a proof uses sequential (weak or strong) compactness, usually one of the three following scenarios applies:

1. The use of sequential compactness (in the strongly compact case) can be replaced by Heine-Borel compactness and so can be reduced to the case of an axiom Δ either by using the binary König's lemma WKL or a—more easily applicable—nonstandard uniform boundedness principle Σ_1^0-UB which follows from a nonstandard axiom $F \in \Delta$ by means of quantifier-free choice (see [Kohlenbach, 1996, 1999] and also [2008b, ch.12]). 'Nonstandard' here refers to the fact that the resp. principles do not hold in the full set-theoretic model. Using a generalized uniform boundedness principle $\exists\text{-UB}^X$ for the type X (see [Kohlenbach, 2006] and [2008b, 17.7.–17.8], for the bounded metric case, and [Günzel and Kohlenbach, 2016] for the normed case) the latter is sometimes even possible in cases where one uses the sequential *weak* compactness e.g. of bounded, closed and convex subsets C of an abstract Hilbert space. Then, however, there is no classically correct principle such as WKL which would imply the Heine-Borel compactness of C (since C is not Heine-Borel compact unless X is finite dimensional) but one *has* to use a strong nonstandard uniform boundedness principle such as $\exists\text{-UB}^X$ which, nevertheless, can be eliminated from the verification of the extracted bound and which does not contribute to the complexity of the bound (see Theorem 3.5 in [Kohlenbach, 2006] or Theorem 17.101 in [2008b]). This principle can also be seen as a version of the bounded collection principle used in the bounded functional interpretation of theories with abstract types (see [Engrácia, 2009] where bounded functional interpretation is used to establish proof-theoretic conservation results for bounded collection).
Uniform boundedness has been used implicitly (and subsequently eliminated) in the unwindings of proofs of theorems of Browder, Wittmann and Yamada (based on a sequential weak compactness argument) in [Kohlenbach, 2011] (for proofs of the theorems of Browder and Wittmann) and [Körnlein, 2016a, Körnlein, 2016b] (for the theorem of Yamada). In both cases, the elimination of the sequential weak compactness argument leaves no contribution to the final bound at all which is of the simple primitive recursive form (+) above (the need for the primitive recursion does not come from the sequential compactness argument but from a projection argument discussed below). Recently, [Ferreira et al., 2018] made the hidden use of nonstandard uniform boundedness on which these proof minings were based explicit

by formulating a general 'macro' which follows from uniform boundedness ($\exists$-UB^X) as well as from bounded collection (in the sense of [Engrácia, 2009]) and can be used to formalize (when adapted suitably to the situations at hand) proofs of the resp. theorems of Browder, Wittmann and a special case of Yamada's theorem due to Bauschke which no longer use sequential weak compactness. Then bounded functional interpretation is applied in [Ferreira *et al.*, 2018] to the resulting proofs to extract bounds similar to those previously obtained in [Kohlenbach, 2011] and—in the more general context of Yamada's theorem—in [Körnlein, 2016b].

Just as the use of sequential weak compactness in the original proof does not contribute at all to the final bounds extraced in [Kohlenbach, 2011] and [Körnlein, 2016b] (which are verified without that use), also the use of Banach limits in a proof analyzed in [Kohlenbach and Leuştean, 2012] in the end turned out to have no contribution to the extracted bound although, in general, such a use *could* contribute very significantly via the comprehension functional ($\exists^2$).

2. Compactness can be avoided altogether by making the original convergence proof constructive once the assumptions are appropriately uniformized. A typical instance of this is the recent unwinding ([Kohlenbach *et al.*, 2018b]) of a noneffective proof for a convergence theorem in the context of the classical Lion-Man game in [López-Acedo, to appear]. Here, using a compactness assumption on a uniquely geodesic space satisfying the so-called betweenness property, it is shown by a nested use of sequential compactness that the lion eventually gets arbitrary close to the man (ε-capture). As it turns out, once the unique geodesic and the betweenness properties are upgraded to 'uniform' versions of these properties (with appropriate moduli so that these properties have the logical form admissible in the logical metatheorems) which—in the presence of compactness—are equivalent to the nonuniform ones, one can avoid the use of sequential compactness in the convergence proof. Moreover, existing metatheorems guarantee the extractability of an explicit rate of convergence (depending on these moduli). This not only provides an effective quantitative version of the original convergence proof but also a vast generalization of the convergence statement itself since now instead of compactness only these uniformized properties need to be assumed. This e.g. applies to all bounded convex subsets of uniformly convex Banach spaces or of CAT(κ)-spaces (of suitably small diameter) and in both cases the respective moduli can

easily be computed (and are low degree polynomials for L^p-spaces and CAT(κ)-spaces).

3. Sequential compactness is eliminated by arithmetizing the original proof. Already in [Kohlenbach, 1998] we showed that the use of (fixed sequences of instances of) sequential compactness in the form of the convergence principle for bounded monotone sequences of reals, the Bolzano-Weierstraß principle for bounded sequences in $\mathbb{R}^n$ and the Arzelà-Ascoli lemma can replaced by arithmetical principles (provable by Σ^0_1-induction which—by [Kohlenbach, 2000]—is optimal) in proofs of $\forall\exists$-statements if the deductive context does not allow for non-quantifier-free instances of induction which use the results of sequential compactness as parameters nor the iteration of the latter. Similarly, the use of (fixed sequences of instances of) the existence of the limsup of bounded sequences in $\mathbb{R}$ can be reduced to Π^0_2-induction (which—by [Kohlenbach, 2000]—again is optimal). The approach is based on an 'elimination of monotone Skolem functions' procedure which in [Kohlenbach, 2008b, (17.9)] is shown to apply also in the presence of abstract spaces X. This explains why in many proofs, despite of the use of sequential compactness, this 'arithmetization' of the proof results in primitive recursive bounds. E.g. this is the case in the elimination of a Bolzano-Weierstraß argument for abstract compact metric spaces in [Kohlenbach, 2005b] and—much extended—in [Kohlenbach et al., 2018a]. Here the use of compactness is not eliminated, in fact the bounds depend on a given modulus of total boundedness of the metric space, but the computational strength of its sequential form just causes a simple primitive recursive contribution to the rate of metastability which is of the form (+) discussed above. Note that in general already the full principle of convergence for monotone sequences in $[0,1]$ is equivalent to arithmetical comprehension ([Simpson, 1999]) whose Gödel functional interpretation cannot be solved in Gödel's T (but requires bar recursion $B_{0,1}$ of lowest type; see [Kohlenbach, 2008b]). Sometimes, the principle of monotone convergence cannot only be reduced to its primitive recursively bounded metastable version but—in the course of the analysis of a proof of a Π^0_2-theorem—to an instance of this metastable version with a very simple counterfunction g (e.g. a constant function $g \equiv k$). This is the reason why the analyses of proofs that originally used the monotone convergence principle as carried out in [Nicolae, 2013, (Thm.4.4)] as well as

in [Kohlenbach et al., 2017, (Thm.3.1)] resulted in simple exponential rates of convergence (in both cases the theorems in question state that a sequence of positive reals decreases towards 0 which is in Π_2^0).

Projection arguments are usually treated by first replacing them in a given proof by arithmetical ε-weakenings. E.g. the existence of the metric projection of $x \in X$ onto the (closed and convex) fixed point set $Fix(T)$ of a nonexpansive mapping T in a Hilbert space X, which is used in the aforementioned proofs of theorems of Browder, Wittmann and Yamada, can usually be replaced (see [Kohlenbach, 2011] and [Körnlein, 2016b] and—similarly—the recent [Ferreira et al., 2018]) by its arithmetical version:

$$\forall \varepsilon > 0 \, \exists y \in Fix(T) \, \forall z \in Fix(T) \, (\|x - y\| \leq \|x - z\| + \varepsilon).$$

While the proof of the existence of the actual projection uses countable choice ([Kohlenbach, 2010]), the arithmetic version can be proved by induction and its quantitative version has a simple primitive recursive bound.

Browder's theorem (proved independently also by Halpern [1967]) states that for a Hilbert space X, a bounded closed and convex subset $C \subseteq X$ and a nonexpansive mapping $T : C \to C$ the path (x_t) of resolvents

$$x_t = tTx_t + (1-t)x, \quad t \in (0, 1), x \in C,$$

strongly converges for $t \to 1^-$ and, in fact, to the metric projection of x onto $Fix(T)$.

In the important paper [Reich, 1980], Browder's theorem was for the first time generalized from Hilbert spaces to more general Banach spaces X such as uniformly smooth spaces. Even for L^p-spaces (other than L^2) this was new. Unless X is a Hilbert space, the path (x_t) never converges to the metric projection of x but to the so-called sunny nonexpansive retraction of x onto $Fix(T)$. In general, nonexpansive retractions onto closed, bounded and convex sets C are known to exist only in special situations, e.g. when C is the fixed point set of a nonexpansive mapping as was shown by Bruck using Zorn's lemma ([Bruck, 1970, 1973a, 1973b]). The only more 'constructive' approach to the existence of (unique sunny) nonexpansive retractions in this situation in fact stems from Reich's theorem. So here one cannot rely on a quantitative version of the existence of ε-versions of sunny nonexpansive retractions to analyze Reich's proofs but has to directly analyze the strong convergence of (x_t). This is done in [Kohlenbach and Sipoş, 2018] (where, in fact, a variant of Reich's proof due to [Morales, 1990] is analyzed). [1990] uses the existence of infima

of the function
$$F(y) := \limsup_{n \to \infty} \|x_{t_n} - y\|^2$$
where (t_n) is a sequence in $(0, 1)$ which converges to 1. In [Kohlenbach and Sipoş, 2018]—using as an additional hypothesis that X is uniformly convex (in addition to being uniformly smooth, which still covers all L^p-spaces for $1 < p < \infty$)—a modulus of uniqueness for the infimum is constructed and used to replace the (unique) point where the infimum is attained by ε-infima. This in turn makes it possible to replace the existence of F as an object (which requires uniform arithmetical comprehension ($\exists^2$)) by

$$\forall y \in C \, \exists z \in \mathbb{R} \, (z = \limsup_{n \to \infty} \|x_{t_n} - y\|^2),$$

where '$z = \limsup_{n \to \infty} \|x_{t_n} - y\|^2$' $\in \Pi_3^0$, which only requires ordinary arithmetical comprehension.

Finally, in the whole proof even the use of limsup's is replaced by ε-limsup's whose existence is equivalent to Π_2^0-induction (that Π_2^0-IA suffices is straightforward; for the converse one has to adapt the proof of Theorem 6.1 in [Kohlenbach, 2000]). Hence—by [Parsons, 1972]—the functional interpretation (combined with negative translation) of the existence of approximate limsup's can be carried out in the fragment T_1 of Gödel's T (which only has the primitive recursive recursors R_0 and R_1). Finally, the existence of the resulting ε-infimum problem can be solved by functionals in T_2. A detailed analysis of the latter solution shows that in the concrete application at hand, the use of type-2 primitive recursion actually reduces to a type-1 primitive recursion resulting in a final rate of metastability for the Cauchy property of (x_{t_n}) which is definable in T_1 (see [Kohlenbach and Sipoş, 2018]). It seems likely that a further analysis of the (very complicated detailed structure of) this bound—in the line of Lemma 4 in [Parsons, 1970]—might show that it actually is definable already in T_0. This then would leave the rate extracted for Baillon's nonlinear ergodic theorem in [Kohlenbach, 2012] (which is definable in T but—as it stands—not in T_0) as is the only bound extracted so far which is not primitive recursive in the ordinary sense of Kleene (i.e. in T_0). Among all the bounds extracted which are definable in T_0, the one obtained in [Safarik, 2012] is the only one which does not have the simple form (+).

The situation discussed can be summarized as follows:

- While mostly uniform *classes* of metric and normed spaces X are used (as atoms) quantification over $\mathbb{N}$ is needed in all applications (already to speak e.g. about the convergence of sequences in X and rates of convergence or metastability), i.e. one does not have model-theoretic tameness and Gödelian or H. Friedman-type phenomena could occur in principle.

- *Empirical fact 1:* all of the rates of asymptotic regularity extracted so far are either polynomial or simple exponential in the basic data (for [Kohlenbach, 2001] and [Kohlenbach and Leuştean, 2003] this holds for constant $\lambda_k = \lambda$ only). This also applies to the moduli of uniqueness extracted in best approximation theory (see [Kohlenbach, 2008b, ch.16], for a survey).

- *Empirical fact 2:* among all the numerous rates of metastability extracted only 2 so far are not primitive recursive as it stands (but definable in Gödel's T and so $\alpha < \varepsilon_0$-recursive). With one exception, the primitive recursive bounds extracted all have a the simple form (+) discussed above. Usually, the primitive iteration involved in (+) can be shown to be necessary by establishing that the Cauchy property of the respective sequence (already in simple cases such as $C := [0, 1]$ and computable mappings $T : [0, 1] \to [0, 1]$) implies Σ_1^0-induction.

- In contrast to model-theoretic tameness, to *detect* proof-theoretic tameness *requires* to actually carry out a *proof analysis* in each individual case.

- *Geometric properties* such as uniform convexity and smoothness etc. are usually *more important than complicated inductions* (see, however, the discussion of [Kohlenbach and Sipoş, 2018] above). The proof-theoretic tameness of the axiomatization of these properties amounts to having a simple (if not trivial) monotone functional interpretation in suitable moduli $\omega : \mathbb{N} \to \mathbb{N}$ which quantitatively witness these properties. See e.g. [Bačák and Kohlenbach, 2018] for converting prima facie noneffective proofs into explicit low-complexity transformations from certain moduli (e.g. of the uniform convexity of the given space) into others (e.g. of uniform continuity for the proximal mappings in uniformly convex spaces).

- The use of *uniform boundedness* (amplifying the already implicitly present uniformity in proofs and corresponding to the use of ultraproducts in continuous or positive bounded logic) does not contribute to the growth of extractable bounds.

- General proof-theoretic *logical metatheorems* play an important guiding role in finding promising applications.

Let us finally mention another important aspect of the proof-theoretic tameness of proofs in nonlinear analysis which has been observed over the past 20 years. Even if a proof does not use noneffective set-theoretic existence principles or complicated inductions, so that the extracted bounds are guaranteed to be of low complexity and even polynomials, the depth of the nesting of the basic functions, e.g. the degree of the polynomial, in general depends superexponentially on the quantifier-complexity of the formulas used in cuts (modus ponens). This already can happen in plain logic or logic augmented by purely universal axioms. Related to this, the bounds extracted by functional interpretation, which make use of the typed λ-calculus, could require superexponentially (in the degree of the highest type use) many β-reduction steps to compute their normal form. In practice, however, the normalization has never been a problem and usually is almost trivial (except for [Kohlenbach and Sipoş, 2018] where things are more involved) and the logical nestings of the basic functions are of very low depth. So not only is the use of mathematical principles typically tame proof-theoretically but even that of first-order logic.

References

[Ariza-Ruiz et al., 2015] D. Ariza-Ruiz, G. López-Acedo, and A. Nicolae. The asymptotic behavior of the composition of firmly nonexpansive mappings. *Journal of Optimization Theory and Applications*, 167:409–429, 2015.

[Bačák and Kohlenbach, 2018] M. Bačák and U. Kohlenbach. On proximal mappings with Young functions in uniformly convex Banach spaces. *Journal of Convex Analysis*, 25:1291–1318, 2018.

[Baldwin, 2018] J.T. Baldwin. *Model Theory and the Philosophy of Mathematical Practice. Formalization without Foundationalism*. Cambridge University Press, Cambridge, 2018. xi+352 pages.

[Bauschke, 2003] H.H. Bauschke. The composition of projections onto closed convex sets in Hilbert space is asymptotically regular. *Proceedings of the American Mathematical Society*, 131:141–146, 2003.

[Bruck, 1970] R.E. Bruck. Nonexpansive retracts of Banach spaces. *Bulletin of the American Mathematical Society*, 76:384–386, 1970.

[Bruck, 1973a] R.E. Bruck. Properties and fixed-point sets of nonexpansive mappings in Banach spaces. *Transactions of the American Mathematical Society*, 179:251–262, 1973.

[Bruck, 1973b] R.E. Bruck. Nonexpansive projections on subsets of Banach spaces. *Pacific Journal of Mathematics*, 47:341–355, 1973.

[Cho, 2016] S. Cho. A variant of continuous logic and applications to fixed point theory. https://arXiv:1610.05397. Preprint 2016.

[Engrácia, 2009] P. Engrácia. *Proof-Theoretical Studies on the Bounded Functional Interpretation*. PhD Thesis, Universidade de Lisboa, 2009.

[Ferreira et al., 2018] F. Ferreira, L. Leuştean, and P. Pinto. On the removal of weak compactness arguments in proof mining. https://arXiv:1810.01508. Preprint 2018.
[Friedman, 1976] H. Friedman. Systems of second-order arithmetic with restricted induction (abstract). *Journal of Symbolic Logic*, 41:558–559, 1976.
[Gerhardy and Kohlenbach, 2008] P. Gerhardy and U. Kohlenbach. General logical metatheorems for functional analysis. *Transactions of the American Mathematical Society*, 360:2615–2660, 2008.
[Günzel and Kohlenbach, 2016] D. Günzel and U. Kohlenbach. Logical metatheorems for abstract spaces axiomatized in positive bounded logic. *Advances in Mathematics*, 290:503–551, 2016.
[Halpern, 1967] B. Halpern. Fixed points of nonexpanding maps. *Bulletin of the American Mathematical Society*, 73:957–961, 1967.
[Kohlenbach, 1996] U. Kohlenbach. Mathematically strong subsystems of analysis with low rate of growth of provably recursive functionals. *Archive for Mathematical Logic*, 36:31–71, 1996.
[Kohlenbach, 1998] U. Kohlenbach. Arithmetizing proofs in analysis. In J.M. Larrazabal, D. Lascar, and G. Mints, editors, *Logic Colloquium 96, Springer Lecture Notes in Logic 12*, pages 115–158. Springer, Berlin-Heidelberg, 1998.
[Kohlenbach, 1999] U. Kohlenbach. The use of a logical principle of uniform boundedness in analysis. In A. Cantini, E. Casari, P. Minari, editors, *Logic and Foundations of Mathematics*, Synthese Library, volume 280, pages 93–106. Kluwer Academic Publishers, 1999.
[Kohlenbach, 2000] U. Kohlenbach. Things that can and things that cannot be done in PRA. *Annals of Pure and Applied Logic*, 102:223–245, 2000.
[Kohlenbach, 2001] U. Kohlenbach. A quantitative version of a theorem due to Borwein-Reich-Shafrir. *Numerical Functional Analysis and Optimization*, 22:641–656, 2001.
[Kohlenbach, 2005a] U. Kohlenbach. Some logical metatheorems with applications in functional analysis. *Transactions of the American Mathematical Society*, 357(1):89–128, 2005.
[Kohlenbach, 2005b] U. Kohlenbach. Some computational aspects of metric fixed point theory. *Nonlinear Analysis*, 61:823–837, 2005.
[Kohlenbach, 2006] U. Kohlenbach. A logical uniform boundedness principle for abstract metric and hyperbolic spaces. In G. Mints, R. de Queiroz, editors, *Proceedings of WoLLIC*, volume 165 of *Electronic Notes in Theoretical Computer Science*, pages 81–93, 2006.
[Kohlenbach, 2008a] U. Kohlenbach. Effective uniform bounds from proofs in abstract functional analysis. In B. Cooper, B. Loewe, and A. Sorbi, editors, *New Computational Paradigms: Changing Conceptions of What is Computable*, pages 223–258. Springer-Verlag, New York, 2008.
[Kohlenbach, 2008b] U. Kohlenbach. *Applied Proof Theory: Proof Interpretations and their Use in Mathematics*. Springer Monographs in Mathematics. Springer-Verlag, Heidelberg-Berlin, 2008. xx+536 pages.
[Kohlenbach, 2010] U. Kohlenbach. On the logical analysis of proofs based on nonseparable Hilbert space theory. In S. Feferman and W. Sieg, editors, *Proofs, Categories and Computations. Essays in Honor of Grigori Mints*, pages 131–143. College Publications, London, 2010.
[Kohlenbach, 2011] U. Kohlenbach. On quantitative versions of theorems due to F.E. Browder and R. Wittmann. *Advances in Mathematics*, 226:2764–2795, 2011.
[Kohlenbach, 2012] U. Kohlenbach. A uniform quantitative form of sequential weak compactness and Baillon's nonlinear ergodic theorem. *Communications in Contemporary Mathematics*, 14:20 pages, 2012.

[Kohlenbach, 2016] U. Kohlenbach. On the quantitative asymptotic behavior of strongly nonexpansive mappings in Banach and geodesic spaces. *Israel Journal of Mathematics*, 216:215–246, 2016.

[Kohlenbach, 2017] U. Kohlenbach. Recent progress in proof mining in nonlinear analysis. *IFCoLog Journal of Logics and their Applications*, 10(4):3357–3406, 2017.

[Kohlenbach, 2019] U. Kohlenbach. A polynomial rate of asymptotic regularity for compositions of projections in Hilbert space. *Foundations of Computational Mathematics*, 19:83–99, 2019.

[Kohlenbach, forthcoming] U. Kohlenbach. Kreisel's "shift of emphasis" and contemporary proof mining. Chapter for forthcoming book *Intuitionism, Computation, and Proof: Selected Themes from the Research of G. Kreisel*.

[Kohlenbach and Leuştean, 2003] U. Kohlenbach and L. Leuştean. Mann iterates of directionally nonexpansive mappings in hyperbolic spaces. *Abstract and Applied Analysis*, 2003(8):449–477, 2003.

[Kohlenbach and Leuştean, 2009] U. Kohlenbach and L. Leuştean. A quantitative mean ergodic theorem for uniformly convex Banach spaces. *Ergodic Theory & Dynamical Systems*, 29:1907–1915, 2009.

[Kohlenbach and Leuştean, 2012] U. Kohlenbach and L. Leuştean. Effective metastability of Halpern iterates in CAT(0) spaces. *Advances in Mathematics*, 231:2526–2556, 2012. Addendum in *Advances in Mathematics*, 250:650–651, 2014.

[Kohlenbach et al., 2017] U. Kohlenbach, G. López-Acedo, and A. Nicolae. Quantitative asymptotic regularity for the composition of two mappings. *Optimization*, 8:1291–1299, 2017.

[Kohlenbach et al., 2018a] U. Kohlenbach, L. Leuştean, and A. Nicolae. Quantitative results of Fejér monotone sequences. *Communications in Contemporary Mathematics*, 20(2):42 pages, 2018. DOI: 10.1142/S0219199717500158.

[Kohlenbach et al., 2018b] U. Kohlenbach, G. López-Acedo, and A. Nicolae. A quantitative analysis of the 'Lion-Man' game. https://arXiv:1806.04496. Preprint 2018. Submitted.

[Kohlenbach and Safarik, 2014] U. Kohlenbach and P. Safarik. Fluctuations, effective learnability and metastability in analysis. *Annals of Pure and Applied Logic*, 165:266–304, 2014.

[Kohlenbach and Sipoş, 2018] U. Kohlenbach and A. Sipoş. The finitary content of sunny nonexpansive retractions. https://arXiv:1812.04940. Submitted.

[Körnlein, 2016a] D. Körnlein. *Quantitative Analysis of Iterative Algorithms in Fixed Point Theory and Convex Optimization*. PhD Thesis, TU Darmstadt, 2016.

[Körnlein, 2016b] D. Körnlein. Quantitative strong convergence for the hybrid steepest descent method. https://arXiv:1610.00517.

[Kreisel, 1951] G. Kreisel. On the interpretation of non-finitist proofs, part I. *Journal of Symbolic Logic*, 16:241–267, 1951.

[Kreisel, 1952] G. Kreisel. On the interpretation of non-finitist proofs, part II: Interpretation of number theory, applications. *Journal of Symbolic Logic*, 17:43–58, 1952.

[Kreisel, 1985a] G. Kreisel. Logical foundations, a lingering malaise. Unpublished Manuscript. Stanford University 1985.

[Kreisel, 1985b] G. Kreisel. Mathematical Logic: Tool and Object Lesson for Science. *Synthese*, 62:139–151, 1985.

[Leuştean and Nicolae, 2016] L. Leuştean and A. Nicolae. Effective results on nonlinear ergodic averages in CAT(k) spaces. *Ergodic Theory & Dynamical Systems*, 36:2580–2601, 2016.

[Leuştean *et al.*, 2018] L. Leuştean, A. Nicolae, and A. Sipoş. An abstract proximal point algorithm. *Journal of Global Optimization*, 72:553–577, 2018.

[López-Acedo, to appear] G. López-Acedo, A. Nicolae, and B. Piątek. "Lion-Man" and the fixed point property. To appear in *Geometriae Dedicata*.

[Morales, 1990] C.H. Morales. Strong convergence theorems for pseudo-contractive mappings in Banach space. *Houston Journal of Mathematics*, 16:549–558, 1990.

[Neumann, 2015] E. Neumann. Computational problems in metric fixed point theory and their Weihrauch degrees. *Logical Methods in Computer Science*, 11:44 pages, 2015.

[Nicolae, 2013] A. Nicolae. Asymptotic behavior of averaged and firmly nonexpansive mappings in geodesic spaces. *Nonlinear Analysis*, 87:102–115, 2013.

[Parsons, 1970] C. Parsons. On a number theoretic choice schema and its relation to induction. In A. Kino, J. Myhill, and R.E. Vesley, editors, *Intuitionism and Proof Theory, Proceedings of the Summer Conference at Buffalo, N.Y.*, volume 60 of *Studies in Logic and the Foundations of Mathematics*, pages 459–473. North Holland, Amsterdam-London, 1970.

[Parsons, 1972] C. Parsons. On n-quantifier induction. *Journal of Symbolic Logic*, 37:466–482, 1972.

[Reich, 1980] S. Reich. Strong convergence theorems for resolvents of accretive operators in Banach spaces. *Journal of Mathematical Analysis and Applications*, 75:287–292, 1980.

[Safarik, 2012] P. Safarik. A quantitative nonlinear strong ergodic theorem for Hilbert spaces. *Journal of Mathematical Analysis and Applications*, 391:26–37, 2012.

[Simpson, 1999] S.G. Simpson. *Subsystems of Second Order Arithmetic, Perspectives in Mathematical Logic*. Springer-Verlag, Berlin-Heidelberg, 1999. xiv+445 pages.

[Tao, 2008a] T. Tao. Soft analysis, hard analysis, and the finite convergence principle. In T. Tao, *Structure and Randomness: Pages from Year One of a Mathematical Blog*, pages 17–29. American Mathematical Society, Providence, Rhode Island, 2008. Essay posted 23 May 2007. 298 pages.

[Tao, 2008b] T. Tao. Norm convergence of multiple ergodic averages for commuting transformations. *Ergodic Theory & Dynamical Systems*, 28:657–688, 2008.

4
Kreisel and Gödel

CHARLES PARSONS[1]

Before I get into my main subject, I would like to say a little about my own relations with Georg Kreisel. I first heard of his work during a year I spent right after college at the University of Cambridge. A neighbor in the student residence at King's College where I was living came upon Kreisel's early papers "On the interpretation of non-finitist proofs" [17] (1951/52). Neither of us had the background to read them, although I had studied mathematics and a little logic as an undergraduate at Harvard. But I knew a little about the Hilbert program and was intrigued by Kreisel's idea of obtaining constructively provable information from consistency proofs.

I returned to Harvard after that year, and early in my second year I had a discussion with Burton Dreben, who had just joined the faculty, about ideas for a dissertation. He said straight off that I should do mathematical logic. I was not attracted by his own project on the decision problem for first-order logic, but proof theory was a natural direction in which to go. Although no one at Harvard before Dreben had much interest in constructivism,[2] I had been stimulated by another teacher in Cambridge to begin reading papers by Brouwer. As for Kreisel, it quickly emerged that to read him I first had to study Hilbert and Bernays, *Grundlagen der Mathematik* [14] (1934/39), and the better part of the academic year 1956–57 was devoted to working through that book, although I did begin to read Kreisel.[3]

As you know Kreisel was during this time (1955–57) at the Institute for Advanced Study. In the summer of 1957 I managed to spend some time at the legendary Summer Institute of Symbolic Logic at Cornell University. Kreisel was there, and I met him and, unless memory deceives me, I had one conversation with him.

As those of you who knew him know, Kreisel was a prolific letter writer. Fairly early I became one of his correspondents, and even while working on my dissertation I received some letters from him. They contained many sug-

---

[1] Edited reading text of a talk to the conference "Kreisel's interests," University of Salzburg, 13 August 2018.

[2] Hao Wang was an exception. But his appointment had ended in 1956. Although he returned in 1961 (with an appointment not in philosophy), I had already finished by then.

[3] The second edition (1968/70) appeared only after my thesis was completed.

gestions and some results, not all of which I understood well. However, Kreisel did not correct me on some points where a proof theorist ought to have intervened, and it was only rather late that I got them right. He was not a detail man. Dreben, my supervisor, though warm, encouraging, and a careful reader and aggressive critic, did not really have enough knowledge of proof theory, and no one else in the Boston area did either. (Hilary Putnam, who probably did, did not come to the area until after I finished.) My accomplishments in that field were modest, and I do not reproach either Dreben or Kreisel personally for their not being greater.

I have not mentioned my other two teachers, Hao Wang and W.V. Quine. Wang left Harvard after my first year in the graduate program, but in that year I took a seminar in which, among other things, he lectured on the consistency proof of *Ackermann* [1] (1940). After he left, he wrote a letter with suggestions for my dissertation that proved very helpful but did not go into any detail. Quine had great influence on me, but more in philosophy than in mathematical logic. He did not know proof theory, and his influence on the dissertation, less than it should have been, was more in the direction of greater precision, accuracy and clarity in exposition.[4]

1

My paper is based largely on the correspondence of Gödel and Kreisel. I never met or saw Gödel, and I did not hear about how they related to each other from others who had visited the IAS. The Gödel papers held by the Princeton University Library (although they belong to the IAS) contain many letters from Kreisel, from 1955 to 1972. I doubt that there are many further letters from Kreisel to Gödel, given Gödel's tendency to keep things. Later, I will have something to say about the last letter from Kreisel in the collection. As regards letters from Gödel to Kreisel, there are only a few copies in the Gödel papers. Since the letters of both are mostly handwritten, it is not surprising that on the whole Gödel did not keep copies of his own letters to Kreisel. There are somewhat more letters from Gödel to Kreisel in the Stanford archives. All are dated between 1959 and 1969. With the exception of a few more official letters, all the letters are in German, and all but a few are handwritten.[5] I have

---

[4] I have written elsewhere on Quine's influence during my student days. See Parsons [21] (2002), as well as the Preface to Parsons [20] (1983).

[5] All translations of letters in German or quotations from them are my own.

not investigated whether other archives might hold letters of Gödel to Kreisel.[6]

In 1955, Gödel invited Kreisel, then a lecturer at the University of Reading in England, to spend the following academic year at the Institute for Advanced Study. Gödel's letter shows that Kreisel had applied for a year's membership in the IAS. It seems clear that Kreisel had not previously visited the United States, and Gödel had never left that country after he emigrated there in 1940. So it is virtually certain that Gödel and Kreisel had not met before. Kreisel spent a second year at the Institute before returning to England.[7] It was during his stay there that he took up with Verena Dyson, the wife of the IAS physicist Freeman Dyson.[8] Although Verena Huber-Dyson reports that the relationship lasted until 1959, there is no indication that it had any substantial effect on Kreisel's relations with Gödel. On the other hand, Gödel did worry about how Kreisel's relations with other Institute members, Dyson in particular, would go during later shorter visits that Kreisel would make to the Institute. At one point he suggests that Kreisel should avoid social interactions with others at the Institute during one visit.

I'm not sure I have straight Kreisel's movements during the period that concerns me. He was away from Reading during the two years at IAS, 1955–57, back there for 1957–58, but at Stanford for 1958–59. He writes that Reading did not give him leave for the Stanford visit, and that he would resign. But one letter, of 31 December 1959, was sent from Reading University. It appears that he did manage to come back for 1959–60. 1960–61 was the first of several visits to Paris. 1961–62 is not clear, but he was in Paris for at least some of the time. In 1962–63 he seems to have been mostly at Stanford but spent some time in Paris; it was in the following summer that the well-known seminar on the foundations of analysis took place at Stanford, with contributions by Kreisel, W.W. Tait, W.A. Howard, and Solomon Feferman, and presence of others, such as Rohit Parikh, Dana Scott, myself, and several students. A letter to Kreisel from J. Robert Oppenheimer, Director of the Institute, invites him there for 1963–64. It was in 1964 that Kreisel was appointed to tenure at Stanford, with the understanding that he would be less than full-time. He seems to have been back at Stanford in April. He was in Paris during 1964–65; because

---

[6]There are Kreisel materials at the University of Konstanz in Germany, mostly still uncatalogued. A note from an archivist there to Mark van Atten states that that archive contains little correspondence and none from Gödel. Thanks to van Atten for this information.

[7]This is confirmed by a letter of Gödel of 27 October 1956 recommending Kreisel (unsuccessfully) for a position as Reader in Mathematical Logic at Oxford.

[8]For an often touching account of this very intense relationship, see Huber-Dyson 1996. She writes that their relationship ended (to all appearances amicably) around the end of Kreisel's 1958–59 visit to Stanford.

of his presence I arranged to be there from February to June 1965. I have not tried to trace Kreisel's movements later. His home base was Stanford up until his premature retirement in 1985. Since by then Gödel was dead, there could not have been any interaction between them afterward.

Most of you, at least those who know German, will have observed how formal German-speaking people were with professional colleagues even well into the post-war period. There are, however, nuances. I did not obtain copies of all of Kreisel's letters in the Gödel papers, but into the 1960s he was using the salutation "Sehr geehrter Herr Professor!" and closing "Ihr sehr ergebener." The last letter, from 1972, begins "Lieber Herr Gödel!" and closes simply "Ihr G. Kreisel." Gödel, already in 1958, uses "Lieber Dr. Kreisel!" and after Kreisel's appointment with tenure at Stanford, "Lieber Professor Kreisel!" The close is always simply "Ihr Kurt Gödel." Perhaps a little of more informal American ways infiltrated his German.

Speaking generally about their correspondence: Two themes seem to dominate: problems in mathematical logic, including results that Kreisel reports, and possible visits of Kreisel to the IAS, where Gödel sometimes writes of complications. Since it is the period after 1957 that is at issue, one might expect Freeman Dyson to be hostile to Kreisel and maybe to oppose visits to the IAS by him. This does not appear to have happened. In a letter of 12 February 1962 Gödel writes

> Since I hear that you are on bad terms with several faculty members, in particular with Dyson, because of the relationship that you are supposed to have had with his former wife, you would have to be careful in your social relations and your statements, perhaps avoid "social contacts" in the framework of the Institute as far as possible.

Then Gödel suggests that Kreisel might take up this question with Oppenheimer. He does not suggest that Dyson would oppose his being invited, although it is very likely that he would not want to see Kreisel socially. In the letter just quoted, Gödel immediately turns to discussion of Spector's consistency proof for analysis.

Of course there are other nonmathematical themes. The career problems of some others are mentioned, particularly Tait and Spector.

A frequent nonmathematical topic in their correspondence is their health. Gödel's concern with his health is well known. He discovered a kindred spirit in Kreisel, and both mention health in their letters. This issue surely created mutual sympathy between them.

Gödel's first letter, inviting him to the Institute for Advanced Study in 1955, is in English.[9] There was little correspondence from the period 1955–57, when Kreisel was in residence there. However, there is one undated letter of Kreisel on IAS letterhead. I think it likely that it dates from 1957, near the end of Kreisel's first stay at the IAS. Mainly, it comments on Gödel's functional interpretation of first-order arithmetic, which was not at that time published. Kreisel acted later to persuade Gödel to publish it in the issue of *Dialectica* in 1958 that was dedicated to Paul Bernays.[10] He also publicized it at the Summer Institute of Symbolic Logic at Cornell University in the summer of 1957 and at the conference on Constructivity in Mathematics in Amsterdam the following September.

Surprisingly, there is almost no discussion of philosophy in the letters I have seen, in spite of the fact that it is frequently said about Gödel that in the 1940s he turned his principal energies to philosophy. Indeed, his principal later mathematical achievement, his rotating solutions of the equations of general relativity, is generally thought to have been originally stimulated by reflection on Kant's view of space and time, a pillar of his transcendental idealism. In the letters I have seen, even the mathematics of general relativity scarcely comes up. It is mentioned only in one of Gödel's last letters to Kreisel.

2

I now turn to the mathematics. On 27 August 1956, Kreisel reports that the young Russian mathematician A.A. Mucnik had obtained the solution of Post's problem independently of Richard Friedberg. He says that Friedberg wanted to publish his work quickly. Kreisel suggests the *Proceedings of the National Academy of Sciences*, which Gödel could arrange.[11]

---

[9]He is addressed as Mr. Kreisel. That was quite correct, since Kreisel did not have a doctorate. He obtained an Sc.D. in 1963, a doctorate that was given by British universities on the basis of published work. Gödel did address him as Dr. Kreisel before he obtained this degree, and, after 1964, as Professor Kreisel. This counts as unusually formal, since most logicians (including such a minor actor as me) addressed him simply as Kreisel. Verena Dyson writes that she addressed him thus even at the height of their relationship. She says that the only person she knew of who called him Georg was Kleene.

[10]See Gödel [7] (1958).

[11]Friedberg's paper was in fact published there as Friedberg [5] (1957), in the 1957 volume of the NAS Proceedings. But Mucnik's paper is in a volume of *Doklady akademii nauk* SSSR dated 1956.

In January 1957 Kreisel reports on the early work of Solomon Feferman, taking off from Turing's idea of ordinal logics.[12] He also says he is looking for a better job in England, although he would prefer the United States. (On this more below.)

On 30 September 1957, Kreisel has returned to Reading and wrote to Gödel from there. He wrote that he had felt unwell during the whole time after he left Princeton and had a high fever from a cold after he arrived in England (four days before), which was, however, now better. He has attended the conference Constructivity in Mathematics in Amsterdam, where he had discussions with Paul Bernays and Arend Heyting. He mentions various details about intuitionistic logic. For example he writes

> Both Bernays and I asked him [Heyting], if one can iterate the highly impredicative concept of intuitionistic logic so readily (Bernays accepts $A \to B$), but Heyting did not see any problem there.

At the time of this correspondence, the consistency of analysis (second-order arithmetic) was a kind of holy grail of proof theory. Kreisel writes that Bernays found Gödel's idea for a proof of the consistency of analysis very promising. Kreisel adds that he (Bernays) sees the interest of such a proof more in the detailed analysis of formal proofs than in the greater evidence that it might yield. (What Gödel's idea was is not explained in the correspondence, and I have not encountered an explanation of it elsewhere. Probably it was preserved only in shorthand notes.)

Kreisel was beginning to be concerned with the completeness of intuitionisitic first-order logic. He writes that a negated prenex formula is derivable in intuitionistic logic if and only if it is derivable in classical logic. But he expresses doubt as to whether completeness of intuitionistic logic obtains intuitionistically for such formulae.

A letter of December 1957 undertakes a long analysis of E.W. Beth's proof of the completeness of intuitionistic first-order logic.[13] The letter also sketches ideas for a proof of the consistency of classical analysis from the assumption that intuitionistic analysis is ω-consistent. It's not entirely clear what formulation of intuitionistic analysis he has in mind.

On 27 February 1958, Kreisel urges Gödel to contribute to the Bernays Festschrift, which Gödel did (see [7]). On 8 March, he mentions the possibility

---

[12] See Feferman [3] (1962).

[13] The letter is probably a prototype of Verena Huber-Dyson and G. Kreisel, "Analysis of Beth's semantic construction of intuitionistic logic," Technical Report, Stanford University, 1959.

of an offer of a visiting appointment at Stanford. There had been a possibility of his spending the summer at IBM, but on 20 March he reports that this has fallen through but the Stanford possibility is still open. On the 20th (probably before receiving Kreisel's letter of the 19th), Gödel writes commenting that Kreisel had apparently not felt entirely well since returning to England. He then notes:

> It would naturally please me very much if you would join the circle of American logicians on a permanent basis. I have already written in this sense to Stanford.[14]

With respect to the question of the completeness of intuitionistic logic for negations of prenex formulae, Gödel says that to prove it one needs a stronger version of the intuitionistic fan theorem, not asserted by intuitionists.

In the summer of 1958 Kreisel was occupied with the proposed analysis of finitism that he was to present to the International Congress of Mathematicians in Edinburgh in the early fall. There was some discussion of issues concerning this with Gödel. A letter of September 24 belongs to this discussion, but I have not been able to understand it, in part because I do not have Gödel's side. On 5 November, he tells Gödel of a theorem of Feferman, that all arithmetically definable models of arithmetic are isomorphic; i.e., that there is no arithmetically definable non-standard model. He also writes about problems concerning his lectures on constructive mathematics, where he is clearly developing his theory of constructions, including the idea (which in the long run was not so successful) that the predicate 'c is a construction proving the proposition p' should be decidable.

In a letter of 2 January 1959, Kreisel reports contact with a "young man in the philosophy department," evidently William Tait.

On 22 February 1959, he reports that the mathematics department at Stanford was prepared to offer him a position but could only cover half of one; the philosophy department did not have the money for the other half. So on April 14 he says that he will return to England for a year. As noted above, it seems that he had not resigned from Reading yet, and they were willing to have him back. In July Kreisel expresses the hope to stop in Princeton in August. In reply, Gödel expresses pleasure about this. He also apologizes for not answering earlier letters. He mentions his own poor health in the spring and a death in his family. This would have been that of his mother-in-law, who had been

---

[14] A footnote adds "not only for personal reasons." [The flag for this footnote is not visible on my copy of the letter. But I think this is the most likely location.]

living with the Gödels since 1956.[15] On 24 August, on the way back to Europe, Kreisel mentions for the first time the work of Clifford Spector. It comes up in another note a couple of days later. In 1961, he will be much occupied with Spector's consistency proof of analysis using bar recursive functionals of arbitrary finite type.

On 31 December, he writes that he has been offered a visiting professorship in Paris but remarks that he will find it tiring to speak French all the time. On the other hand, he hopes that he will be able to learn how much mathematicians use impredicative methods. (I don't know of his ever publishing on this subject.)

On 28 February 1960, he says it is settled that he will be in Paris the next year but hopes to spend some time at the IAS before the semester begins there (November 1).

On 20 March 1960, he mentions Kleene's work on intuitionistic analysis. He says that he would be interested in a refutation of Takeuti's conjecture (that cuts can be eliminated from formal proofs in second-order logic). Instead, this was *proved* a few years later by Tait and generalized to higher types by Takahashi and Prawitz.

On 1 June, he writes that he has proved that one cannot prove the weak completeness of intuitionistic first-order logic by intuitionistic means.

On 20 December, writing from Paris, he reports that he is suffering from stomach trouble, which he attributes to eating in restaurants. He also reports that the local logician Daniel Lacombe had been suspended after participating in a demonstration against the Algerian War. The suspension was lifted after a not very long interval.

This provoked a comment from Gödel on 17 March 1961:

> What you wrote about Lacombe and the conditions at the University of Paris is unfortunate, but still it is worth recognizing that Lacombe was soon restored to his job. In France democracy is too deeply rooted for it to be destroyed overnight as in Germany.

Kreisel had also reported on a visit to East Berlin. He reports that Karl Schröter, the leader of logic in East Germany at the time, had a cut-free version of modal predicate logic. More interestingly he observed that Markov, a leader of Russian constructivism at the time, was present. Kreisel remarks that he was "very dogmatic" about his version of constructivity. Kreisel talks of his lecture course on classical logic in Paris, which was proceeding model-theoretically.

---

[15] See Dawson ([2] (1997), p.213).

He remarks that he has obtained Herbrand's theorem but that Herbrand's rules do not fit well into his framework. He claims to have a simpler proof of Beth's definability theorem.

Gödel's letter of 17 March 1961 does not respond to the mathematical content of Kreisel's. He writes that he is trying to arrange for Kreisel to visit the Institute in the next academic year and then turns to Spector's work. "I have the impression that he has hopes for a much more constructive consistency proof for analysis, whose existence is [in Gödel's view] very questionable. It would be lovely if one could get by with computable functionals of transfinite type (say up to ε_0). One can, I believe, really not demand more." Spector has told him that Tait has an idea for a proof by means of the substitution method, up to a very large constructive ordinal and asks if Kreisel has heard of it. Then he remarks:

> I have the impression that mathematicians in general are negatively disposed toward partial carrying out of the Hilbert program, although only through that (effort) can these questions really be clarified. It is possible that Spector felt this attitude of mathematicians (for example at his lecture in Michigan) and that it had a depressing effect on him.[16]

Gödel then mentions what was probably his first encounter with Abraham Robinson's non-standard analysis, which over time was to make a great impression on him.

Kreisel, replying on 22 March, is very critical of Tait's idea for proving the consistency of analysis by the substitution method.[17] Commenting on what Gödel wrote about Spector's professional difficulties, he says,

> It seems to me very premature to in the present state of things to expect of mathematicians (or even logicians!) a lot of understanding of the intuitionistic interpretation of classical analysis. Only when the basic idea of this interpretation has been actually applied to simple concrete cases and constructively shaped in proofs present in a concrete sense will one find understanding among mathematicians.

On 3 August, however, he reacts to the news of Spector's death. He had received Spector's manuscript the evening before and begun to work with it

---

[16] In fact, Spector was offered and accepted a position at Michigan but died before he could take it up.

[17] However, in a letter of early April, he praises Tait's eventually published work on the substitution method. Cf. Tait [24] (1965).

but says he was "paralyzed" by the news of his death.[18]

In a letter from the spring of 1961, Kreisel had mentioned Solomon Feferman's work on predicativity. He says there is some difficulty about Feferman's promotion to tenure.[19]

At some point in 1961 Kreisel was proposed for a position, probably as a Senior Research Fellow, at All Souls College, Oxford. There is a letter from Michael Dummett to Gödel, dated October 6, asking him for a letter about Kreisel. I did not find a copy or draft of a letter by Gödel in his papers at Princeton.[20]

Gödel wrote in January 1962 about a possible visit by Kreisel to the Institute. He says that "suddenly objections were made, which according to my previous information I absolutely could not expect." He can, however, offer Kreisel a place there as his assistant. He says that since members have no duties their stipends are free of income tax, but since assistants officially have obligations to the relevant professor, their stipend is taxable. He says that he would not impose any duties on Kreisel. He wants Kreisel to come for three semesters and thinks (on these not so attractive terms) that he can bring it about.

Non-Americans and even younger Americans might be baffled by this. Up until a law passed in 1986, research fellowships were not taxable, at least up to a limit of $ 300 a month. Gödel writes as if the exemption were unlimited, and it may well have been in 1962.[21]

Kreisel had written of health problems and complained that he had not found a good doctor. Gödel mentions a friend from his student days, Marcel Natkin, who had lived in Paris for thirty years, and says he is about to write to him and evidently ask for recommendations.

---

[18] See Spector [23] (1962). The paper as published was edited by Kreisel. But see the postscript by Gödel, p.27. I had attended the symposium in New York where Spector spoke on this work, and as I recall we had a pleasant conversation afterward. It did not occur to me that Spector had only a few months to live.

[19] No such difficulty is mentioned in Feferman's (truncated) autobiography. See Jäger and Sieg ([16] (2017), pp. xxxvi–xxxvii). Feferman's classic paper "Systems of predicative analysis" [4] (1964), is based on an invited lecture to the ASL in January 1963.

[20] I recall that rumor had it that he was not elected because he was thought not sufficiently "clubbable." More recently it was suggested to me that anti-Semitism played a role. I never learned about the truth or falsity of these rumors.

[21] I don't recall my stipend as a Junior Fellow of the Society of Fellows at Harvard (1958–61) as having been taxed. It may have been more than $ 300 a month, but not a lot more.

On 12 February Gödel writes that he is happy to learn that the invitation to the Institute for a longer time is not important for Kreisel. He mentions as important for proof theory the characterization of finitary mathematics and the question whether "Brouwer's principle of defining ordinal numbers" (presumably the bar theorem or principles related to it), together with functionals of finite type is sufficient for the consistency of analysis.

> There are remarkably few logicians who are interested in questions of this kind, and I am afraid that in another position your work would also go in another direction.

This is a somewhat odd remark, since at the time Kreisel did not have a steady position, but both may have anticipated that he would have one at Stanford.

Gödel repeats the offer for Kreisel to come to the Institute as his "assistant" and mentions the social difficulty (see above). He turns to Spector's proof and says an *Ergänzung* (enlargement or completion) is necessary. He doesn't consider this a correction. "The main point is in any case, that Spector's proof is correct in all details."[22]

On 9 March, Gödel writes again, reporting that after discussion with Oppenheimer, that shorter or longer visits by Kreisel can always be arranged, provided that the costs are covered by his assistantship funds. Before July 1, he can come only for a two-week visit. Gödel says he can't communicate the reasons for the resistance of the faculty (evidently to a more generous arrangement). He says that none of his colleagues has said he is on bad terms with Kreisel, but for some there is a lack of good will toward him. Gödel has heard that Dyson has no objection to a stay by Kreisel, but he (Gödel) has no personal contact with him. The topic turns to health, and Gödel mentions the doctor in Paris recommended by his old friend Marcel Natkin.

Returning to mathematics, he argues that the vicious circle principle does not hold "in its full extent" for constructive mathematics. He refers to pp.133–34 of his Russell paper (Gödel [6] (1944))[23] but adds remarks about intuitionism that were not in that paper. He remarks that in intuitionism the concept of totality does not occur at all, because there quantifiers are to be interpreted intensionally.

Kreisel replied on March 20. He remarks that he is pleased that the objection to inviting him to the Institute did not come from Dyson. He would like to

---

[22]These remarks suggest that Kreisel had not finished the editing of Spector's paper, but that is not explicitly said. It seems to contradict the remark on the first page of the printed paper that it was received by the editor of the symposium proceedings on September 23, 1961.

[23]This corresponds to pp.125–26 of Gödel [9] (1990).

come to the Institute before June but says that would conflict with his duties in Paris. But he says that he might be able to come for two or three weeks during June.

Kreisel reports on a paper he has received from Tait, which contains a new proof of the no-counter-example interpretation of arithmetic "without detour through the ε-substitution method" and obtains the first ε-number in a "completely natural way," which also yields ω-consistency. But he complains that the result is concealed under a lot of *Kleinarbeit*.

He then turns to politics, remarking on the revolt of the OAS (*Organisation de l'armée secrète*). He has the impression that "both the governments and the population in France and among the Arabs would now prefer accepting a compromise to the domination of the OAS." At the end he promises that he will not provoke Gödel's colleagues and says he is convinced that there will be no conflicts. Gödel's reply of 11 April expresses pleasure at the prospect of Kreisel's visit in June and infers from Kreisel's letter that the latter's health is tolerable. He considers that the general theorem underlying the substitution method for arithmetic and transfinite recursion rests is "very elegant." He makes other remarks about Tait's work, in particular that Tait now believes that the substitution method is not promising for analysis.[24]

Commenting on Kleene's paper on recursive functionals of arbitrary finite type, Gödel notes that its concept of primitive recursive function of lowest type agrees with the older one.

> However, that means that the functions definable in the system T of my Dialectica paper are not all primitive recursive. That makes it doubtful to me, whether this concept of "primitive recursive" is the right one.

In a footnote he remarks that they are the provably recursive functions of PA, i.e. those ordinal recursive in ordinals less than ε_0. In a P.S., Gödel remarks that the mood of the faculty is beginning to change in a way favorable to him (Kreisel).

It appears that Kreisel did not visit the Institute in June. On May 29, he wrote expressing worry whether his visa would allow him to receive pay from the IAS. Gödel wrote on 30 July, saying that there is no hurry about his decision about the post as assistant. He asks how things stand with the characterization of finitary and predicative mathematics and briefly comments on Takeuti 1961. On 1 August, Kreisel writes sketching the idea of his later paper with Troelstra on choice sequences.

---

[24]If my memory is correct, Tait was at the Institute at this time.

Kreisel had remarked that Tait's promotion at Stanford was under consideration and asked Gödel to write a letter. It's not clear from my record whether Gödel did this. But on 8 August, Kreisel writes that Tait has been rejected, evidently on the initiative of John Myhill:

> Myhill is very critically disposed toward Tait and the direction of his work and considers him a poor teacher. Moreover, he would very much like to bring Putnam to Stanford and to hold a position open for that. For that reason he pressed strongly (in my absence) for the dismissal of Tait, and in fact the philosophy department voted for it. I have maintained (with Tait's and Myhill's agreement) that one should to begin with extend his appointment for one or two years, and that should be decided at the beginning of October.[25]

It's not clear whether the latter option left the question of Tait's tenure open, but other indications, including remarks to me by Tait, indicate that it did.

On 17 September, Gödel writes of an official invitation to the Institute for Kreisel with the title Research Associate. Whether he actually spent 1962–63 at the Institute is unclear from the correspondence. However, he wrote from Stanford on 22 September expressing pleasure about the IAS prospect (also with comments on Takeuti's work), but on 26 December he wrote from Paris speaking of a strenuous term at Stanford. He remarks, "I find, happily, that logic instruction in Paris is going really well." On 9 January 1963, still in Paris, he reports attending a photography exhibit of Natkin.

On 15 April he writes from Stanford with a preliminary report on Paul Cohen's work, saying that he has the independence of AC and has an idea about CH.

It appears that Kreisel's visit to the Institute was postponed until the next year. There is a letter from Oppenheimer formally offering him an appointment for 1963–64, with a stipend of $ 10,000. However, in the spring Kreisel appears to be at Stanford.

On 28 January 1964, Patrick Suppes asked Gödel for a letter about Kreisel. But he says, "We will only be able to have him with us part of each academic year." Gödel wrote the recommendation on 14 February.

On 15 April 1964, Kreisel writes to Gödel asking for a letter for Tait and thanks him for it on 10 June. It appears, however, that Tait was rejected and

---

[25] According to usual practice, the negative votes mentioned would still have left Tait with an appointment for 1962–63. So that proposal would extend it until 1965, which is in fact when Tait left Stanford.

that Kreisel did not support him at this point.[26] That he would take this stance does not appear in the correspondence. On 10 June, thanking Gödel for the letter, he remarks:

> Indeed, I found your appraisal of the situation very interesting; for example, I can't remember having mentioned Howard's weak point in your presence. In conversation with him one is struck by this point, but—so it seemed to me—less in his contribution to the Report.[27]

Kreisel did not explain more fully what Howard's weak point was.

About the rejection of Tait, Kreisel wrote that it was decided not to change Tait's job into a permanent one.

> The main reasons were an internal affair. In any case the decision was free of all personal prejudices; also Tait's achievements were rated highly.

I doubt that Tait himself was convinced of this. In 1965 Kreisel wrote that Tait was angry with him.

3

My examination of Kreisel's later letters to Gödel has been somewhat limited. Furthermore, after 1969 replies by Gödel are not available in archives known to me. On 18 June 1964, Gödel suggests a later date for Kreisel to visit the Institute but suggests that he come for a couple of days in the summer and then for one or two weeks in the fall. Kreisel had said something relevant to the consistency of analysis, and Gödel remarks that if his idea holds up then there are good prospects for a consistency proof for analysis. He mentions Hausdorff's "Pantachieproblem," which he had mentioned to Paul Cohen and suggested that he try either to prove the existence of an increasing sequence of sequences of numbers of type ω_1 that majorizes every arbitrary ω-sequence of numbers, or to prove its unprovability, or to show its compatibility with $\neg$CH.[28]

Kreisel had written on 8 August 1964 about an injury to his arm. On 27 October Gödel replied, expressing concern that "the injury to your arm was really not so trivial as it initially appeared to be." He then comments on a result

---

[26] I am indebted here to Tait. Myhill had left Stanford and was not involved at this point.

[27] Report of a seminar on the foundations of analysis, Stanford University, 1963. The reference is to the logician W.A. Howard, who had participated in the seminar in 1963.

[28] See Gödel's letter to Paul Cohen of 22 January 1964, (Gödel [11] (2003), pp.383–84).

of Tait. He makes comments on choice sequences in the role of arguments and on the predicate calculus in systems Takeuti works with.

On 22 February 1965, Kreisel points out that if one could choose a distinct real number for every countable ordinal, then it would follow without AC that $\aleph_1 \leq 2 \exp \aleph_0$. He suggests that Gödel should write more about the Dialectica interpretation. Somewhat later, Gödel did undertake this in the notes that he added to an English version of his paper, but he never sent back the proofs for the revised paper to be printed. (See Gödel ([9] (1990), especially pp.).)

In a letter of 6 July 1965, Gödel turns to set theory. Kreisel had given an example where an existential quantifier is not constructible. Gödel says the example is "very pretty" and is the first one [of this kind] where one has a proof. He comments that the compatibility of GCH with the existence of a measurable cardinal has not been proved, and he is inclined to believe that they are incompatible. (This conjecture turned out to be wrong.) He has heard that Frederick Rowbottom has proved that if there is a measurable cardinal, then aleph-1 in the model of constructible sets is weakly compact.

On 4 April 1966, Kreisel reports on Dana Scott's method of obtaining Cohen's results by means of Boolean-valued models. He reports that Kripke and Martin have shown it consistent that there are Σ_4^1 and Π_4^1 sets of integers that are not constructible. (That such sets exist, given large cardinals, follows from the results of Solovay [22] (1967).)

In a letter of 18 August 1966, Gödel finds the visa difficulties that Kreisel is having regrettable and asks if they will prevent him from teaching at Stanford in the summer of 1965 and 1966. He hopes that Kreisel can come to IAS for a week in the summer. This must have been arranged, because he writes on 10 August that an apartment has been arranged for him. He makes a number of remarks about set theory, infinitary logic, and Takeuti's work. On 19 October, he hopes to see Kreisel in Princeton. He regrets that Kreisel's health has not been good but hopes that going to Europe will do him good. "You have there, obviously, also your accustomed physician."

On 9 March 1967, Gödel thanks Kreisel for a letter and says he has no objection to his mentioning the history of his *Dialectica* paper in a lecture. He asks whether Feferman's development of a constructive theory of ordinals of higher types made progress. (It's not clear whether it's the theory or the ordinals that are said to be of higher type. It seems more likely to be the former.)

He asks whether a working out of Kreisel's Edinburgh lecture (on a characterization of finitism) has appeared or will appear. He then raises a technical difficulty about it.

He asks Kreisel to send greetings from him and his wife to a certain Raubitschek, evidently a mathematician who had been at Princeton and then moved to Stanford.

On 20 December 1967, Gödel thanks Kreisel for two letters and some reprints. But then he says, "Unfortunately, my being occupied with philosophy has prevented me from replying sooner." He regrets that a visit of Kreisel in August did not take place but would want to see him if he should come into the neighborhood.

Readers of *The Magic Mountain* would be amused by Gödel's closing paragraph:

> The view of the sanatorium in Davos that you sent is *really* beautiful. I hope that your visit there was not caused by illness, or that your state of health is now again satisfactory.

In 1968, Kreisel was evidently nominating Gödel to be a Fellow of the Royal Society. Gödel's letter of 1 May 1968 comments on Kreisel's draft of a statement he had prepared to recommend him as a candidate. In his letter of 9 June, it's not clear whether he is still commenting on that statement. He says that when he talked of "elementary combinatorial" proofs, he had in mind what Hilbert called finit, as explained in Hilbert [13] (1926). He remarks:

> Since here all abstractions and psychological reflections are excluded, and Hilbert obviously was also not thinking of a "training" or other "conditioning" of intuition, but only wanted to apply what is as immediately evident to everyone as simple complete induction, it thus seems clear to me, that one in essence may apply only those inductive inferences and definitions that are immediately clear without proof through mere representation of the ordinal number or reduction procedure.

On 25 July 1969, Gödel wrote a letter to Kreisel which ended with a remark that may have contributed to Kreisel's apparently ending their relationship three years later:

> I would be very happy to see you again from time to time. However, I must unfortunately give up occupying myself more closely with the works of other logicians, if I want, at the age of 63, to achieve anything essential in philosophy or with respect to the true power of the continuum.

This was the last letter of Gödel to Kreisel preserved in the Stanford collection, with the exception of a brief and cordial Christmas card sent at the end of that year. In 1970 Gödel was ill and during this time exhibited paranoid symptoms. Kreisel continued to write even into 1972, mentioning mathematical results, but I don't have evidence that Gödel replied.[29] However, the effects on others of Gödel's illness and mental state must have reached Kreisel. Gödel recovered and during the next couple of years had an active relationship with Hao Wang and played a role in the last revisions and sending in for printing of Wang's book 1974.[30]

Kreisel is known for instances in which he became intimately involved intellectually with other logicians, carried on this relationship for some years, and then broke off the relationship. It seems that Kurt Gödel was not immune. That is shown by Kreisel's letter of 1 October, 1972, of which I have added the full text in the original and in my translation. I do not know of Gödel's having replied. It is very probable that they never saw each other again.

However, Kreisel was not done with Gödel. After Gödel's death in 1978 he wrote a "biographical memoir" for the Royal Society (see Kreisel [18] (1980)) and at least one other article on Gödel's work (Kreisel [19] (1983)).

References

[1] Ackermann, Wilhelm, 1940. Zur Widerspruchsfreiheit der Zahlentheorie. *Mathematische Annalen* 117, 162–194.

[2] Dawson, John W., Jr., 1997. *Logical Dilemmas: The Life and Work of Kurt Gödel*. Wellesley, Mass.: A.K. Peters.

[3] Feferman, Solomon, 1962. Transfinite recursive progressions of axiomatic theories. *The Journal of Symbolic Logic* 27, 259–316.

[4] Feferman, Solomon, 1964. Systems of predicative analysis. *The Journal of Symbolic Logic* 29, 1–30.

[5] Friedberg, Richard, 1957. Two recursively enumerable sets of incomparable degrees of unsolvability (Solution of Post's problem, 1944). *Proceedings of the National Academy of Sciences USA* 43, 236–38.

[6] Gödel, Kurt, 1944. Russell's mathematical logic. In P.A. Schilpp (ed.), *The Philosophy of Bertrand Russell*, pp.123–154. Evanston, Ill.: Northwestern University. Reprinted in Gödel 1990.

---

[29] For that reason I am puzzled by the reference in Kreisel's letter of 1 October 1972 to "our conversations."

[30] See the introductory note to Wang's correspondence with Gödel in Gödel [12] (2003a).

[7] Gödel, Kurt, 1958. Über eine bisher noch nicht benützte Erweiterung des finiten Standpunktes. *Dialectica* 12, 280–287. Reprinted with translation in Gödel 1990.

[8] Gödel, Kurt, 1972. On an extension of finitary mathematics which has not yet been used. In Gödel 1990, pp.271–280.

[9] Gödel, Kurt, 1990. *Collected Works, volume II, Publications 1938–1974*. Solomon Feferman et al., eds. New York and Oxford: Oxford University Press.

[10] Gödel, Kurt, 1995. *Collected Works, volume III, Unpublished Essays and Lectures*. Solomon Feferman et al., eds. New York and Oxford: Oxford University Press.

[11] Gödel, Kurt, 2003. *Collected Works, volume IV: Correspondence A–G*. Solomon Feferman et al., eds. Oxford: Clarendon Press.

[12] Gödel, Kurt, 2003a. *Collected Works, volume V, Correspondence H–Z*. Solomon Feferman et al., eds. Oxford: Clarendon Press.

[13] Hilbert, David, 1926. Über das Unendliche. *Mathematische Annalen* 95, 161–190.

[14] Hilbert, David, and Paul Bernays, 1934/39. *Grundlagen der Mathematik*. 2 vols. Berlin: Springer.

[15] Huber-Dyson, Verena, 1996. Thoughts on the occasion of Kreisel's 70th birthday. In Piergiorgio Odifreddi (ed.), Kreiseliana: *About and around Georg Kreisel*, pp.51–73. Wellesley, Mass.: A.K. Peters.

[16] Jäger, Gerhard, and Wilfried Sieg (eds.), 2017. *Feferman on Foundations: Logic, Mathematics, Philosophy*. Dordrecht: Springer.

[17] Kreisel, G., 1951/52. On the interpretation of non-finitist proofs. *The Journal of Symbolic Logic* 16, 241–267; 17, 43–58.

[18] Kreisel, G., 1980. Kurt Gödel, 28 October 1906 – 14 January 1978. *Biographical Memoirs of Fellows of the Royal Society* 26, 148–224.

[19] Kreisel, G., 1983. Gödel's excursions into intuitionistic logic. In Paul Weingartner and Leopold Schmetterer (eds.), *Gödel Remembered*, pp.65–179. Naples: Bibliopilis.

[20] Parsons, Charles, 1983. *Mathematics in Philosophy: Selected Essays*. Ithaca, N.Y.: Cornell University Press.

[21] Parsons, Charles, 2002. W.V. Quine: A student's-eye view. *Harvard Review of Philosophy* 10, 6–10.

[22] Solovay, Robert M., 1967. A non-constructible Δ_3^1 set of integers. *Transactions of the American Mathematical Society* 127 (1967), 50–75.

[23] Spector, Clifford, 1962. Provably recursive functionals of analysis: A consistency proof of analysis by an extension of principles formulated in current intuitionistic mathematics. In *Recursive Function Theory: Proceedings of Symposia in Pure Mathematics*, vol.5, pp.1–27. Providence: American Mathematical Society.

[24] Tait, W.W., 1965. The substitution method. *The Journal of Symbolic Logic* 30, 175–192.

[25] Takeuti, Gaisi, 1961. Remarks on Cantor's absolute. *Journal of the Mathematical Society of Japan* 13, 197–206.

[26] Wang, Hao, 1974. *From Mathematics to Philosophy*. London: Routledge and Kegan Paul.

Appendix

Kreisel's (apparently) last letter to Gödel (On letterhead of Stanford University, Department of Philosophy) 1.10.72

Lieber Herr Gödel!

Was ich hier zu sagen habe, hätte ich mir sehr lange überlegen können, habe es aber nicht getan. Nachdem ich es nun getan habe, wäre es einfach schäbig – nach einer Bekanntschaft von 17 Jahren – Ihnen diese Überlegungen zu verschweigen.

Seit Ihrer akuten Krise vor ein paar Jahren von krankhaften Mißtrauen (gegenüber Ihnen wohlgesinnten Menschen) finde ich unsere Gespräche unbefriedigend. Ich bin überzeugt, Sie auch, da ja solche Gefühle beinahe immer gegenseitig sind.

Vor allem stören mich alle Zeichen von Mißtrauen. Es wirkt besonders komisch, wenn es sich um Dinge handelt, die ich viel besser kenne als Sie, z.B. die Umstände, unter welchen meine Mitgliedschaft bei der A.M.S. verfallen ist. Da ich Sie selten treffe, traue ich mir selbst keine grossartige Vergleiche zwischen Ihrem Zustand jetzt und vor 17 Jahren zu: Aber folgendes ist mir aufgefallen.

Schon früher störte mich Ihr – mir übertrieben scheinendes – Mißtrauen in Ihren politischen Überlegungen. Ich habe sie höflich gebeten, mit mir nicht darüber zu sprechen: *diesen Wunsch haben Sie erfüllt*. In den letzten Jahren habe ich Sie auch gebeten, nicht immer von Silver-Solovay (SS) zu sprechen (und ich habe, höflich, darauf hingewiesen, daß andere Ihrer Bekannten besser über die SS informiert sind als ich): *diesen Wunsch haben Sie nicht erfüllt*.

Ihr Mißtrauen gegenüber SS könnte[31] auch folgende Rolle spielen. Wenn man keine Lust hat, sich in die Beweise von SS zu vertiefen, aber glaubt verpflichtet zu sein, sich darum zu kümmern, so liefert das Mißtrauen einen objektiven Grund für den Mangel an Interesse! – Wie dem auch sei, habe ich mehrmals ganz explizit erwähnt, daß ich eine gewisse Interessenlosigkeit Ihrerseits (für Dinge, die nicht ganz dicht mit Ihren Interessen aus den 30er Jahren verbunden sind) zu spüren glaubte. Jedenfalls werden Sie bemerkt haben, daß ich in den letzten Jahren – ganz im Gegensatz zu unseren Gesprächen in den ersten 10 Jahren unserer Bekanntschaft – keine Lust hatten, Ihnen neue Ergebnisse in der Logik, weder von anderen noch von mir, zu berichten.

---

[31] Mißtrauen scheint eine erstickende – und für mich sehr unangenehme – Krankheit zu sein.

Wie gesagt, ich behaupte gar nicht, daß sich vieles bei Ihnen in den letzten zehn Jahren geändert hätte. Nach meiner Erfahrung mit vielen Menschen (nicht aufgrund ungeprüfter "Theorien") scheinen die menschlichen Beziehungen eine gewisse natürliche Lebensdauer zu haben. Wenn diese abgelaufen ist, dann fallen einem am anderen Züge auf, die immer da waren, aber keine störende Wirkung hatten. Das ist alles – und neue Beziehungen werden angeknüpft. Gottseidank (was mich anbetrifft), bestehen meine Schwierigkeiten nicht in esse, d.h. wenn ich an unsere Bekanntschaft denke, sondern nur in re. Und natürlich bin ich mir bewußt, daß ich in den ersten 10 Jahren unserer Bekanntschaft viel von Ihnen gelernt habe, und zwar Dinge, die ich schon vorher suchte und nirgends fand. (Wie ich auch irgendwo schon erwähnte, habe ich erst durch Gespräche mit Ihnen den richtigen Zugang zu Ihren Publikationen gefunden.) Dafür bin ich und bleibe ich Ihnen sehr dankbar. Mit den besten Wünschen auch für Ihre Frau

Ihr G. Kreisel[32]

[Translation of the above]

Dear Gödel,

What I have to say here I could have mulled over for a long time, but I have not done that. Now that I have done it, it would have been simply shabby—after an acquaintance of 17 years—to conceal these reflections from you.

Ever since your acute crisis a few years ago of pathological mistrust (towards people well disposed to you), I have found our conversations unsatisfying. I am sure that you do too, since after all such feelings are nearly always reciprocated.

Above all, I am troubled by all signs of mistrust. It feels especially strange when it is about matters that I am much better acquainted with than you are, such as the circumstances in which my membership in the AMS lapsed. Since I seldom meet you, I don't have confidence in any grand comparisons between your state now and seventeen years ago. But the following has occurred to me:

Already earlier I was disturbed by the mistrust—which seemed to me exaggerated—in your political reflections. I asked you politely not to talk about such matters, and you fulfilled this wish. In recent years I have also asked you not to talk all the time about Silver and Solovay (SS), and I have, politely, pointed out that other acquaintances of yours were better informed about the SS than I was. *This wish you have not fulfilled.*

---

[32]This letter is the property of the Institute for Advanced Study; the (handwritten) original is deposited in the Kurt Gödel papers at Princeton University Library. Copyright is presumably held by the estate of Georg Kreisel.

Your mistrust toward SS could also play the following role.[33] If one has no desire to go deeply into the proofs of SS but believes oneself obligated to be concerned with them, then one's mistrust yields an objective ground for the lack of interest!—However that may be, I have several times mentioned explicitly that I thought I detected a certain lack of interest on your part (for things that were not very closely connected with your interests of the 1930s). In any case you will have noticed that I had in recent years—quite in contrast to our conversations in the first ten years of our acquaintance—no inclination to report to you new results in logic, either others' or my own.

As I have said, I do not claim that a lot has changed with you in the last ten years. According to my experience with many people (not on the basis of untested "theories"), human relations seem to have a certain natural life span. When this has run out, then one is struck by certain features of the other that were always there but had not had any disturbing effect. That is all — and new relationships are established.

Thankfully (as concerns me), my difficulties don't consist *in esse*, that is when I think of our acquaintance, but only *in re*. And of course I am aware of the fact that I learned a lot from you in the first ten years of our acquaintance, in particular things that I was already looking for earlier and found nowhere. (As I also already pointed out somewhere, I first found the right mode of access to your publications through conversations with you.) For that, I am and remain very grateful to you.

With best wishes, also to your wife,

Yours, G. Kreisel[34]

Acknowledgement

Letters of Kreisel are copyright by the Estate of Georg Kreisel and are quoted or published by permission of Professor Paul Weingartner. Those used here are located in the Kurt Gödel papers, the Shelby White and Leon Levy Archives Center, Institute for Advanced Study, Princeton, NJ, USA, on deposit at Princeton University, and are quoted or published with their permission.

Letters of Gödel quoted here are copyright by the Institute for Advanced Study and are quoted by permission of the same Shelby White and Leon Levy

---

[33] Mistrust seems to me to be a suffocating—and for me very unpleasant—illness.

[34] Translation by Charles Parsons, revised using suggestions of Jacob Rosen. The original letter is in the Gödel papers and thus is the property of the Institute for Advanced Study. Copyright belongs to the estate of Georg Kreisel.

Archives Center of the Institute. They are held in the Stanford University Archives and are quoted with the additional permission of the Department of Special Collections of the Stanford University Library.

5
Georg Kreisel: Some Biographical Facts

DANIEL ISAACSON

The biographical facts in this paper are from Kreisel's early life, up to 1950, when he became internationally known with the publication of his first paper in mathematical logic ([Kreisel, 1950]). The interest in this period of Kreisel's life is that relatively little has been known about it, even by people who knew him well later, or indeed at the time, which has led to a certain amount of misinformation being published.

1 Austrian beginnings 1923–1939

Georg Kreisel was born 15 September 1923 in Graz, Austria, the first child of Heinrich Kreisel (b. 21 July 1886, d. in or before 1950), a snail merchant (*Schneckenhändler*)[1], and his wife Berta (b. 31 October 1901, d. 8 December 1950). Kreisel's brother Fritz was born on 12 January 1925. The family lived at 26 Hüttenbrennergasse, which was also the address of Heinrich Kreisel's snail business (see [7], pp.131, 395, and 827).

Kreisel [1998] has recorded the following anecdote from his childhood (p.152): "At the age of 5, I was struck by the phenomenon of twinkling stars, and asked my nanny if this was something in the stars or in my eyes. She did not correct me (apparently agreeing that it was one or the other); only in my teens did I learn about effects of the atmosphere, as it were, in-between (which astronauts know without book learning)."

Matthias Baaz reports[2] that Kreisel told him Erwin Schrödinger visited his parents when Schrödinger was professor at the University of Graz, a post he accepted in 1936 ([Heitler, 1961, p.224])[3]. One wonders how this came about. A possible explanation might be that Schrödinger was fond of snails.

Kreisel's parents were Jewish by descent, and though irreligious or at any rate non-observant (snails aren't kosher), the family was immediately sub-

---

[1] See [17] for an account of the return of snails to Austrian cuisine in recent years.

[2] In discussion following the talk in which I presented some of the material in this paper at the Salzburg meeting.

[3] Schrödinger was born and grew up in Vienna. Before coming to Graz he held posts in Jena, Stuttgart, Zurich, Berlin, and Oxford. In 1933 he and Paul Dirac were awarded the Nobel Prize in Physics for their work on quantum mechanics.

jected to Nazi persecution in the aftermath of the annexation of Austria by Nazi Germany (*Anschluss*) on 12 March 1938, and Kreisel and his brother were forced to leave their school in Graz[4]. Their parents sent them to the Jewish Gymnasium in Vienna, "the only secondary school for Non-Aryans in former Austria" ([Kreisel, 1939, p.118]). They were still in Austria on *Kristallnacht*, 9–10 November 1938. The two boys reached England, at the port of Harwich, via the *Kindertransport*, on 12 January 1939. That day was Fritz's fourteenth birthday and Kreisel was fifteen[5,6].

---

[4]Schrödinger, who was not Jewish, was forced to flee Austria because of his forthright opposition to Nazism when he held a post in Berlin ([Heitler, 1961]).

[5]The most extraordinary piece of misinformation published about Kreisel—we will see several others—is by Verena Huber-Dyson in [1996], when she says (p.54) "his parents had been foresighted and affluent enough to send their two boys to England shortly before the *Anschluss*". Verena Dyson had an intimate relationship with Kreisel over the course of several years, including a period of time when they lived together. It seems to me she can only have believed this if Kreisel told it to her. One might conjecture that he did so to forestall being seen by her as having been vulnerable. Kreisel remarks on this misinformation in a letter to Matt Ridley 20 April 2006: "I have only dipped into *Kreiseliana*, and found many errors; for example, I did not leave Austria before the *Anschluss*" [email message from Kenneth Derus 31 August 2018].

[6]From when Kreisel took early retirement from Stanford in 1985 until he was too infirm to travel, in the last few years before his death in 2015, he was of no fixed address, traveling between ad hoc accommodation in Oxford, Cambridge, Salzburg, Vienna, Munich, Zurich, Sils Maria, staying no longer than a year in any one place, and usually only a few weeks or a month. Francis Crick [1996] reports (p.31) "I have urged him to buy a small house somewhere, if only to have a *pied-à-terre*, but to no avail." He kept only a very few possessions in the way stations of his peregrinations, most notably blackout material so that wherever he slept could be made totally dark (these places also needed to be totally silent) plus a few odd books and papers that had come his way which he hadn't yet disposed of, but essentially what he possessed at any given time was what he carried with him. Traveling with all his possessions in this way is reminiscent of when he came from Austria to England.

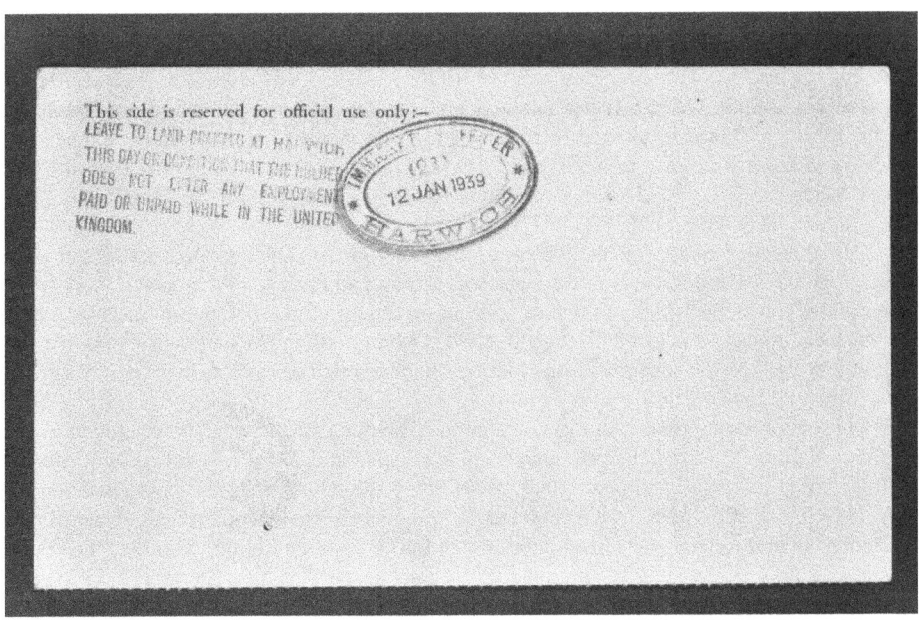

Kreisel's entry into the United Kingdom was granted by this landing card (front and back).

2 Refugee in Dudley Grammar School 1939–1941

Georg and Fritz Kreisel were taken in as refugee students at Dudley Grammar School[7], which was for boys aged from 11 to 18, located in the town of Dudley[8], a medium sized industrial centre in the West Midlands[9]. Fritz entered the school on 14 March 1939, two months after reaching England; Kreisel arrived eight days later. The school had raised money to support three Jewish refugees[10] (the third was 14 year old Kurt Flossman, who had arrived in the school on 30 January of that year).

Kreisel's parents also escaped from Nazi Austria. They "were among a group of Jews who were allowed by the German government to emigrate from the German Reich to Palestine in late summer 1940. They were subsequently detained by the British off the coast of Cyprus. Then they were briefly interned

---

[7]It has sometimes been said that Kreisel attended Dulwich College. The innocent source of this misinformation is Freeman Dyson—see p.101. It has also been said, by Peter Conradi in his biography of Iris Murdoch, that Kreisel was at Rugby ([Conradi, 2001, p.262]); in that same passage, Conradi very plausibly reports Kreisel as having described himself as "the kind of alien England could afford".

[8]Dudley Grammar School was founded in 1562; it had moved to its site on St James's Road in 1898, into a new Victorian Gothic building (described by Pevsner [1974, p.122] as of "flaming red brick"), with additions in 1909, 1926, and 1936 that created a quadrangle (see site plan in [Temple, 1962]). These buildings are still standing—see Google Street View [6], but the school itself ceased to exist in 1975, when it was merged into a mixed comprehensive school, and those buildings, to which others have since been added, are now occupied by St James Academy, part of the Dudley Academies Trust—see [18]. Dudley Grammar School had 424 pupils in 1939 ([Temple, 1962, p.42]).

[9]"Although surrounded by Staffordshire, the borough [of Dudley] was associated with Worcestershire for non-administrative purposes, forming an enclave of the county until 1966, when it was transferred to Staffordshire after an expansion of the borough boundaries. Following local government reorganization in 1974, Dudley took in the boroughs of Halesowen and Stourbridge to form the present-day Metropolitan Borough of Dudley, in the newly formed West Midlands county." ([19])

[10]The school fee in 1930 was £15 per annum ([Raybould, 2010, p.82]) and probably was still that in 1939; the total cost for supporting a refugee student for a year was estimated at £50, including full board and school fees, as set out in the school magazine, *The Dudleian* for July 1940 (issue 107), p.96, which reported on "progress of the School branch of the Mayor of Dudley's Fund for Refugees (itself a branch of Earl Baldwin's Refugee Fund)": "The Fund began with two record Saturday morning collections at the School in December, 1938, and January, 1939, which brought in £7 10s. 0d. and £8 10s. 0d respectively. The total after six months was just over £53, including £10 4s. 6d. from a Whist and Bridge Drive organised by the Parents' Committee on May 16th last year, £1 1s. 5d. from a collection of Parents' Night (June 19th) and grant of £5 from the School Tuck shop. By the end of 1939 it had risen to £85, one of the most useful additions being a collection at the Commemoration Service a year ago which brought in £5 6s. 0d." And so it went.

at the Atlit Detention Camp in Haifa, Palestine, before being deported to Mauritius in December 1940, where they were incarcerated for the remainder of the war" ([Hawkins, 2020, p.188]). The British, as the colonial power, were attempting to block Jewish immigration into Palestine. After the war in Europe had ended, Heinrich and Berta Kreisel were taken back to Palestine, in August 1945 (*loc. cit.*), and they were living there during the creation of the Jewish State of Israel. Kreisel was able to write to his parents while they were held in Mauritius, though how often is not clear, and some at least of those letters were lost (see [9], entry dated 27 March 1945). Kreisel's mother died in Israel on 8 December 1950, aged only 49. We know that Kreisel's father died some time before that since Berta was recorded as a widow at the time of her death ([10]). Matthias Baaz has told me that in a conversation with Kreisel when Baaz was about to leave for a conference in Israel, he asked Kreisel if he had ever been to Israel, to which Kreisel replied that he had been there during the War of Independence (which was 1947–49). This would have been a chance for him to see his mother again, and his father, if he was still alive then. Kreisel's mother left a will, for which Kreisel was executor. The will was probated in London on 17 December 1951, with a probate value of £1150 ([10])[11].

In July 1939, four months after his arrival in Dudley Grammar School, Kreisel published an account in the school magazine, *The Dudleian*, of his experience of *Kristallnacht* in Vienna[12]. By this account, having been grabbed by the SA as he came out of his school, he was turned over to the SS, and after a terrifying night with thousands of other Jewish men and boys who had been rounded up, many beaten and some killed, he was one of only three boys released from incarceration in the morning, while the other prisoners were taken to concentration camps.

This account is at odds on a crucial point with something Kreisel told Freeman Dyson after they became friends in Cambridge. Dyson [2018] wrote in a letter to his parents in February 1943 (p.31), "I spent last evening talking to Georg Kreisel, the Viennese refugee whom I may have mentioned before as being one of the ablest people here. It is remarkable, but I am finding the refugees on the whole the nicest people. In particular, Kreisel is an outstandingly solid character. It makes one feel very small to talk to a person who has been in Dachau concentration camp; Kreisel was there for a fortnight when he was fifteen and spared none of the ghastly details". This raises the question whether

---

[11] According to the Office of National Statistics, the average price of a house in the United Kingdom in 1951 was a bit over £2000 ([16]).

[12] Because this earliest publication by Kreisel is not readily available, I quote it in full as an Appendix to this paper.

Kreisel was suppressing the trauma of having been taken away to Dachau in his *Dudleian* piece, or embellishing the truth in what he told his friend. Online search of the Dachau Concentration Camp Records ([4]), brings up four men with surname 'Kreisel' having been sent to Dachau, none of whom has first name 'Georg' or was born in 1923, which suggests the latter. If this is indeed embellishment, we may surmise that having been one of only three spared the fate of being sent to Dachau or other camps after the intense horrors of the inflicted cruelty experienced and witnessed in *Kristallnacht*, Kreisel would have contemplated that fate with such intensity as to feel its reality.

There is another possibility which, on present knowledge, cannot be ruled out, that not only was Kreisel speaking imaginative rather than literal truth when he told Freeman Dyson he had been sent to Dachau, but that he was doing this also in his "Recollection" of "The Tenth of November" for his school. This possibility would be excluded on finding Georg Kreisel's name in a list of Jews rounded up by the SA and detained by the SS in Vienna on *Kristallnacht*. I have not been able to find SA or SS records online, and lack the expertise to pursue this query in some other way. I hope that this question will be resolved at some point.

Dudley Grammar School was not primarily boarding, but it had a residence, Lingwood House, where Kreisel and his brother lodged. The master of Lingwood House was Dr James Mainwaring[13], who played a significant paternal role for Kreisel during this time, as shown in the following report by Sir John Clapham [1943], Chairman of the Cambridge Refugee Committee (p.4): "George Kreisel, gained a Major Open Scholarship in Mathematics at Trinity College, Cambridge, in December 1941, from Dudley Grammar School. No other money was available and his house-master, who had taken a great interest in the boy ever since his arrival in England, wrote to Lord Baldwin and asked his help [Stanley Baldwin, former Prime Minister, was a key figure in establishing the *Kindertransport*, and at that time Chancellor of Cambridge University; he also lived, in retirement, within 20 miles of Dudley]. Lord Baldwin's Under-Secretary, a Mr. Boss, then offered to help Kreisel during the years at this University with £150 per annum[14] This has made Kreisel self-supporting, as he is able to spend all his holidays with his late house-master."[15] Another

---

[13] A site map of the school published in [Temple, 1962] identifies 12 St James's Road as "'Lingwood'—Home of boarders until 1946 under Dr. and Mrs. J. Mainwaring."

[14] This account clears up an otherwise baffling statement by Francis Crick in [1996, p.27]: "I learnt that Kreisel had come to England as a refugee and had been sponsored by Stanley Baldwin." There could have been no reason for Baldwin to sponsor Kreisel among the thousands of child refugees being accepted into the UK under the *Kindertransport*.

[15] Earlier in this document, in which Kreisel was then cited as a case study, Clapham artic-

indication of James Mainwaring's significant role for Kreisel is in Freeman Dyson's letter to his parents on meeting Kreisel, in which he reports Kreisel as having said that he "picked up his philosophical bent from his housemaster" (see below p.101). A yet further indication of James Mainwaring in a paternal role in Kreisel's life at this time is a note in Kreisel's Jewish Refugees Committee record [8] dated 14 January 1943, in anticipation of Kreisel completing his degree at Cambridge later that year, which says, "Written to Dr Mainwaring asking if he could bring to bear any influence on George to comply more with conventions, as this might affect the type of war work into which he will be directed after taking his exam."

Kreisel had proclaimed his disdain for convention while at Dudley Grammar School, in an article, "The Devil", published in *The Dudleian* in July 1941 ([Kreisel, 1941]). In it he declares that the devil "is a reaction against the conventions and taboos of an age. This fact, of course, only adds to his merit, for only intelligent men can do without taboos and conventions which themselves mechanise human behaviour and avoid the labour of thought. For the devil the need for conventions ceases because of his powerful intelligence. [...] The devil knows no code, he is spontaneous and original." Kreisel evidently identified with the devil. He ends with some strident declarations which, among the conventionally religious, would have been seen as blasphemous: "The devil is the creation of Man, just as God is created in Man's image and not Man in God's. But doubtless the devil is man's pet creation. He is the projection of what is essentially Man: honour, patriotism, martyrdom are invented and destroyed, but Man and the devil remain." Kreisel follows his essay with the Latin adage, "Tempora mutantur, nos et mutamur in illis." ("Times are changed, we also are changed with them.")[16]

---

ulates the financial difficulties faced by bright young refugees capable of going on to university after secondary school but not eligible for a State Scholarship or Bursary, which a British national would be: "The largest Open Scholarship in Cambridge University carries £100 per annum. [...] It is usually considered that a man needs £250 per annum, a woman £200 per annum to live and study in Cambridge. [One wonders why men need 25% more than women, but that's not of concern here.] This estimate is an average one and is meant to cover young British people, who have their own homes to go to in vacation", so with Baldwin's grant and his Major Open Scholarship from Trinity, Kreisel had £250 per annum at Cambridge, which by being able to go back to Lingwood House in Dudley during vacations, gave him sufficient support to live and study in Cambridge as an undergraduate.

[16]Following my talk in the Salzburg meeting, Mark van Atten sent me the following extract from an email message to him from Robert Tragesser on 21 March 2015, recollecting conversations with Kreisel at Stanford: "He'd invite me in for sometimes long, pleasant chats. One thing about his bungalow (which was in what was then a quiet semi-rural area) was that, onto the frame of his front door was placed a devil's head where one would expect a mezuzah, a

Despite all this, *The Dudleian* continued to report his accomplishments proudly and admiringly, both while in the school and during his time at Cambridge. This might reflect, in part, an awareness that they had in their midst a mathematical genius whose subsequent career would bring credit to the school, but I don't think this is a sufficient explanation. Kreisel's genius in combination with his attitude that "intelligent men can do without taboos and conventions" could have engendered hostility, or at best grudging admiration. I conjecture that the fact that this did not happen reflects the importance for Kreisel of the support he received from James Mainwaring, as set out above (p.92). Knowing about this stage of Kreisel's life calls for knowing something about James Mainwaring. I am grateful to Richard Hawkins for providing a significant source ([11]) and other information.

James Mainwaring was a remarkable person. Born in 1892, he came to Dudley Grammar School in 1917 to teach history after an undergraduate degree at Liverpool University and a year teaching in a grammar school in Yorkshire (exempt from military service in World War I by service as a teacher [3]). In 1931 he obtained his M.A. in Education from Birmingham University. In 1937 he became Dudley Grammar School's Sixth Form Master and Head of the Arts Department. In 1940 he was awarded the degree of D.Litt. from the Department of Education at Birmingham University for a thesis on *The Aesthetic Judgment* (part 1 "A critical analysis of the main theories of aesthetics", part 2 "The quality of the aesthetic judgment"). In that year he also published the first volume of a three-volume school textbook *Man and His World* ([Mainwaring, 1940]) (which included his own pen and ink drawings)[17]. There is a clue in this book as to James Mainwaring's steadfast support of Kreisel in the face of his declarations in "The Devil" (see above, p.93) and the elements of his character reflected in them. Richard Hawkins reports, citing [Alcorn, 1999, p.142], that *Man and His World* "caused controversy by suggesting the resurrection of Christ was not necessarily a fact" (email message 8 January 2010). Someone prepared to stand this ground could feel empathy with Kreisel's heroic self-assurance, without having to be heroically self-assured himself.

In addition to teaching history, Dr Mainwaring trained the school choir and the band of the school's cadet corps, and also produced and directed three

---

reminder of God's presence."

[17]Mainwaring appears to have published a number of other books, including *British Social History* (in two volumes), *British History in Strip Pictures*, *Psychology in the Classroom*, and *Teaching Music in Schools*, but I have not been able to obtain a bibliography of his publications, and can't rule out that one or more of these might be by a different James Mainwaring.

Gilbert & Sullivan operas in successive years before the war, and other operas for the Stourbridge Amateur Operatic Society[18]. The website of the Dudley Choral Society records that "The Dudley and District Choral and Orchestral Society was formed in 1945 and gave its first performance in Dudley Town Hall on the 29th December. They performed Handel's Messiah under the baton of the founder Dr Mainwaring." He also throughout this period, with his wife, looked after the pupils of the school who boarded, in his role as housemaster of Lingwood House. He played a key part in setting up the Dudley Grammar School Committee to Aid Refugees from the Nazis. *The Dudleian* March 1939 reports on a Whist Drive to raise money for the school's Refugee Fund: "Forty members and friends enjoyed a most pleasant evening in the Hall, with Mr Mainwaring this time as M.C., and it was largely thanks to his enthusiasm and cheerfulness that the time passed so quickly." Richard Hawkins [2020, fn.18] reports an email to him from a former pupil in the school who characterized James Mainwaring as "charismatic". He left Dudley Grammar School in December 1944 to take up a post at the Teacher Training College in Dudley (later part of the University of Wolverhampton) as tutor in history and music.

Crucial though Mainwaring's support evidently was for Kreisel's success in Dudley Grammar School, I think it must also have depended on the ethos of the school established by its pupil-centred headmaster, the eccentric (by all accounts) but highly effective David Temple, who occupied that post from 1934 to 1962 ([Temple, 1962, p.46]). Trevor Raybould [2010] reports the following impression of him by a pupil who was there during the war years (pp.84–5): "Eager for life, he scurried everywhere. We were all lucky. Not only did Dave keep that school lively and stimulating during the war, coping with staff and materials shortages, while encouraging a remarkably wide range of extra-curricular activities, he also spent countless hours writing letter to old boys in the forces, making sure they knew the school remembered them–and cared. He looked an oddity. His clothes were often–well, rather scruffy, his writing notoriously illegible, and those great ears of his stood out from his head like radar dishes. But he was a lovely man. And he was Dave–the Head." Raybould also reports in detail some of David Temple's other lovable eccentricities ([Raybould, 2010, p.93]).

The school's continuing pride in Kreisel's achievements, and I think also affection, comes out in *The Dudleian* for December 1942, which reported that "Georg Kreisel, already a Scholar and Prizeman of Trinity College, Cambridge, has gained further distinction by winning the Essay Competition, open

---

[18]Music would not have been been a point of contact between Kreisel and Mainwaring. Though acutely sensitive to noise, Kreisel considered himself to be tone deaf.

to the College." When Kreisel completed his Cambridge degree, *The Dudleian* for July 1943 reported: "Overshadowing all former successes gained by Old Boys at Cambridge is the achievement of Georg Kreisel, Mathematical Scholar of Trinity College, who has crowned a brilliant University career by being placed very high in the list of Wranglers[19] of his year. As a result of his performance in the Mathematical Tripos, Part II, Kreisel has just been awarded a Senior Scholarship for post-graduate work. The school congratulates him heartily and has duly celebrated the achievement in the traditional way, by a half-holiday awarded in his honour, and that of L.A. Tromans, the running Blue."

Kreisel's academic prowess in the school is set out in the account of Prize Day, November 25th, 1941 in *The Dudleian* for December 1941, at the end of his time in the school. He is one of five pupils who achieved Northern Universities' Joint Matriculation Board Higher School Certificates (corresponding to what later became A-Levels). This announcement is followed by the recognition, "Distinguished in the Examination:—Applied Mechanics, Pure Mathematics, Physics—G. Kreisel." In the *Valete* (Farewell) section of the *The Dudleian* for March 1942, after he had left the school, these accomplishments are detailed, along with having obtained a First Aid Certificate from the St John's Ambulance Brigade in 1940, and having served as Curator of the School Museum, in which he took a much more proactive stance than his predecessors, as he reported in *The Dudleian* for March 1940: "The Museum, which then had already a history of several years, was actually only a collection of curiosities without any form or planning. Obviously, it was time to look for a theme which the Museum was to serve. The fact that Dudley Grammar School is one of the best-known educational institutions in a district famous for its many industries suggested at once that we should try to provide a series of inspiring exhibits, illustrating the various industries flourishing about here." He then reports having obtained a display on glass making, including glass blowing and painting, from the firm of Stevens and Williams of Brierley Hill. "It is hoped to replace this exhibit soon by another in order to give as complete a picture of the various industries as possible."

Fritz Kreisel was not nearly so academically accomplished as his brother. At the Prize Day, November 25th, 1941, he is among 38 pupils recorded as having obtained their School Certificate (corresponding to later O-levels), and he left school at this stage, aged 16. In the *Valete* section of the *The Dudleian* for December 1941 he is listed as having obtained, in addition to his School

---

[19] A "wrangler" is a student who gains first class honours in the third year exam, Part II, of the Mathematical Tripos in Cambridge University ([21]).

Certificate, a Royal Life Saving Society Bronze Medallion in 1940 and Silver Medallion in 1941. One year after arriving at the school, Fritz published a short story, "The Swing Beam", in *The Dudleian* ([F. Kreisel, 1940]), which though clearly not completely autobiographical, does suggest that he was very unhappy during this time, missing his parents and probably not knowing then whether or not they were still alive. His command of English in it is extraordinarily good. The story itself is desperately sad. It's about a "small, shy boy" named John who "showed his loneliness and homesickness" during the first days of his new school life, who "when he was supposed to be doing his homework [...] would be secretly writing a letter each to 'Dearest Mummy' and 'Dearest Daddy'". He was mercilessly teased and molested by the other boys, who call him 'Roundy'. The "good old Master", described as "Johnny's only friend there", is his only protection (one thinks here of James Mainwaring). John receives news that his mother is very ill. He "made himself believe that if he could manage the swing beam test ['A big horizontal beam tapering off towards the far end [...] supported in such a way that it could swing freely sideways. It was a severe test of skill to keep your balance to walk right along it'] his mother would recover." When he had nearly accomplished this task, the other boys see what he is doing and knock him off. The Master finds him, takes him back to the house, and tries to console him. The Master receives a telegram the next morning and takes Johnny to the station to return home. His mother has died. The boys who had persecuted 'Roundy' then "felt like murderers—murderers of 'Roundy's' mother." Fritz appends a note at the end, "(Based on a German story)".

A non-autobiographical element of "The Swing Beam" is that Fritz had a close friend in the school, John D. B. Edwards (known as JDB), and when Fritz left Dudley Grammar School at 16, JDB's parents invited him to live in their home (see [Hawkins, 2020, p.26]). Fritz went to work in various businesses in Dudley and in Worcester, while planning to enter one of the Fighting Services (entries for 27 July 1942 and 14 April 1944 in his Jewish Refugees Committee record [9]). In April 1944, aged 19, Fritz joined the regular army as a private in the Worcestershire Regiment. In doing so he changed his name to Fred Edwards, taking the surname of his friend and host family. It was standard for Jews joining the British Armed Forces during the Second World War to adopt a completely English-sounding name, to forestall being persecuted as a Jew if captured. Kreisel's Jewish Refugees Committee record ([8]) has a note dated 19 March 1945 saying "letter from Georg to say that he would write to his parents and reassure them that his brother was alright in the army, and that it is the usual practice to change the name as he has done. [His brother] seems to have some objection to writing direct to his parents according to Georg." A note in Fritz's Jewish Refugees Committee record [9] dated 27 March 1945 reports that "Fritz['s] brother George has informed us no need for his father to worry about Fritz change of name—thinks some of his letters explaining must have been lost in transit."

Francis Crick, in his "Personal Recollections" for *Kreiseliana*, notes that "Kreisel seldom spoke about his family background", so that Crick was "surprised when on one occasion he introduced me briefly to his brother" ([Crick, 1996, p.27]) . This may have been during the war, when Crick was in Havant and Kreisel was in London, or it could have been after the war. Richard Hawkins [2020] provides information about the career of Fritz Kreisel/Fred Edwards in the British Army from the point in autumn 1943 when he succeeded in joining the Army (pp.26–7): "After a few weeks of training in Scotland he became a private in the Worcestershire Regiment and was stationed near Dudley. He then fought in north-west Europe. By December 1945 he was serving with the British Army of Occupation in Germany attached to the Intelligence Corps. Fred Edwards later returned to Britain. In September 1947, the month he was naturalized, he was stationed at a German POW camp at Goathurst, near Bridgwater in Somerset, where he was probably an interpreter. He appears to have remained in the armed forces until at least late 1949. [. . .] It is not known what happened to Fred Edwards after 1949."

3 Cambridge undergraduate 1942–1943

In January 1942 Kreisel went up to Trinity College Cambridge to do his undergraduate degree in Mathematics, after two years and nine months at Dudley Grammar School.[20] He was admitted to the College on 13 January [email message from Jonathan Smith 13 April 2016] and matriculated in the University on 28 January [email message from Jacqueline Cox 15 April 2016]. A.S. Besicovitch, about whom Kreisel always spoke warmly, was Kreisel's director of studies at Trinity. Besicovitch was a tough-minded analyst, but when Kreisel told him of his interest in foundations of mathematics, he responded positively and sent him to John Wisdom, a tutorial fellow in philosophy at Trinity, who

---

[20]Dudley Grammar School did not standardly send their pupils to Cambridge or Oxford. The headmaster in 1919 referred to the University of Birmingham as being "naturally the place of higher learning to which boys from this School proceed" ([Raybould, 2010, p.81]). At the same time, the school successfully put forward their best pupils for Oxbridge scholarships, Kreisel being an outstanding example, but there were others, as noted in the tribute to James Mainwaring ([11]).

in turn sent him to Ludwig Wittgenstein, a professorial fellow of the college, who was giving a seminar on foundations of mathematics. "Quite soon Wittgenstein invited me for walks and conversations. This was not entirely odd, since in his (and my) eyes I had at least one advantage over the other participants in the seminar: I did not study philosophy. Be that as it may, in his company (à deux) I had what in current jargon is called an especially positive *Lebensgefühl*," as Kreisel later recalled ("On Some Conversations with Wittgenstein: Recollections and Reflections" [Kreisel, 1989, p.148]).

From September 1941 until March 1943 Wittgenstein was working as a porter at Guy's Hospital in London. During this time he would come back to Cambridge to give a two-hour lecture class on Saturday. He would have a walk with Kreisel on Friday afternoon, and they sometimes saw each other on Saturday evening, after the class. Kreisel reports that "often he sketched, in a few minutes in the course of Friday afternoon walks, the content of his two-hour seminar the next day, or afterwards, on Saturday evening, supplemented or extended that content." ([Klagge and Nordmann, 2003, p.355]). However, their meetings were independent of the lectures Wittgenstein was giving, and instead, at Wittgenstein's suggestion, focused on reading together G.H. Hardy's [1921] *A Course of Pure Mathematics*.

Before the summer vacation of 1942 Wittgenstein loaned Kreisel a copy of *The Blue Book*. Kreisel [1989] reports that (p.151) "In leafing through it I came across the metaphor *family resemblance of concepts*[21]; it pleased me, as I mentioned to him in conversation. But in the course of the summer a malaise developed", which he reports as follows (p.152): "The whole of abstract mathematics is full of families of concepts; for example, the family of groups, of which one has a concept (from the usual definition). But it is not a (tacitly: rewarding) object of research: roughly, what holds for all groups is rarely of use for any particular group. Usually such families have a few broad, simple properties that should not be forgotten. But, at least for common sense, brooding about them is liable to reach the point of diminishing returns quite soon. [...] When I returned Wittgenstein's transcript after the summer vacation I also mentioned my malaise." Wittgenstein was not pleased, and Kreisel recalls him saying something like (p.152) " 'I taught you everything I,'—then he interrupted himself—'you are capable of learning' ". Nonetheless, they continued to meet, and Ray Monk reports that "In 1944—when Kreisel was still only twenty-one—Wittgenstein shocked Rhees by declaring Kreisel to be the most able philosopher he had ever met who was also a mathematician. " 'More

---

[21]Wittgenstein uses the phrase "family likeness" in *The Blue Book* ([Wittgenstein, 1958, pp.17 and 33]).

able than Ramsey?' Rhees asked. 'Ramsey?' replied Wittgenstein, 'Ramsey was a mathematician!' " ([Monk, 1990], p.498). Wittgenstein's high regard for Kreisel is also expressed in a letter to Rhees 18 November 1944: Talking about people who have been coming to his classes, he writes "[...] and a woman Mrs so and so who calls herself Miss Anscombe, who certainly is intelligent, though not of Kreisel's caliber" ([McGuinness, 2012, p.371]).

Kreisel reports that Wittgenstein "once said he was surprised that he could be friends with anybody as irreligious as me. (I was not surprised.) If I had been compelled to be surprised about something of this sort, then at most about the fact that there should have been any talk at all of friendship despite an age difference of nearly 35 years" ([Kreisel, 1989, p.155]). (Wittgenstein was three years younger than Kreisel's father and three years older than James Mainwaring.)

Through contact with Wittgenstein, Kreisel got to know Elizabeth Anscombe, then a postgraduate student at Newnham College, who had come to Cambridge in 1942 from her undergraduate Philosophy degree in Oxford (Greats) to study with Wittgenstein. Later in that decade they were involved together in some madcap episodes ([Conradi, 2001, p.264 and pp.315–6], for which also see [Horner and Rowe, 2015, p.381]). When Wittgenstein died, in 1951, it turned out that he had named Elizabeth Anscombe in his will as one of his three literary executors (the other two being Georg Henrik von Wright and Rush Rhees), and it was she who translated his *Philosophische Untersuchungen*, the one work in his *Nachlass* he had more or less prepared for publication by the time of his death, from Wittgenstein's German into English. In her "Translator's Note" to the *Philosophical Investigations*, 1953, Anscombe lists Kreisel as among those "who either checked the translation or allowed me to consult them about German and Austrian usage or read the translation through and helped me to improve the English" ([Wittgenstein, 1953, p.[v]]).

Soon after arriving at Trinity, Kreisel became friends with Freeman Dyson, a Trinity Mathematics undergraduate of Kreisel's year, who wrote in a letter to his parents 10 February 1942:

> I received a cake from Aunt Margaret last week. I managed to invite a man to eat it with me on Sunday, and I chose a refugee by the name of Chrysel [*Correctly spelled, his name was Georg Kreisel. He played a dramatic role in my story fifteen years later.*] who has come up this term and is a pursuer of "higher thought" in matters mathematical. The tea party was not entirely a failure, and the guest was very appreciative. He talked incessantly about the nature of knowledge. He goes to

some classes held by Professor Wittgenstein of which I was entirely uninformed, but which sound most amusing, though Chrysel like a good German is incapable of taking anything except with deadly earnest. How these Germans take things seriously. Chrysel remarked casually that he had read fifteen philosophical books by Bertrand Russell alone since he came England two and a half years ago, and of course he has read one at least of Kant and Hegel, Descartes and Leibnitz in the originals, three or four of Whitehead, three of Moore, and any number of others. He was at Dulwich School[22] and apparently picked up his philosophical bent there from his housemaster. He is taking mathematics as his main subject and knows a good deal about that too. He lived in Vienna, but I do not know what else he has done. He is a logical positivist. He makes his creed "Everybody else is wrong" rather than "I am right," like Socrates, and is equally annoying if you are not in a good temper, as Socrates's victims usually were not. Fortunately I was in a good temper on Sunday and enjoyed it very much. I shall go to tea with him one of these days ([Dyson, 2018, pp.12–13]).

I asked Professor Dyson if he knew anything about Fritz Kreisel from that time, to which he replied (email message 31 March 2016) that so far as he remembered, Kreisel never mentioned a brother.

Another of Kreisel's close friends while an undergraduate, and continuing afterwards (see [Conradi, 2001, pp.315–6]), was Gabriel Andrew Dirac, stepson of the Cambridge physicist Paul Dirac. Kreisel attended Paul Dirac's lectures, and greatly admired him.

Kreisel sat the Preliminary Examination in Mathematics (Part I) during his second term in Cambridge, Easter Term 1942, and was awarded a first class pass [message from Jacqueline Cox 1 July 2016]. While in his first year as an undergraduate he also pursued his deep interest in foundations of mathematics by mastering the then recently published second volume of *Grundlagen der Mathematik* by David Hilbert and Paul Bernays. Kreisel [1987a] speaks of having been (p.395) "attracted by the logical wit of consistency proofs (which I learnt in 1942 from Hilbert-Bernays Vol.2)". Kreisel's study of Hilbert-Bernays Vol.2 was on his own—no one in Cambridge at that time would have been competent to teach it, and this laid the foundation for the research he pursued for the rest of his life. When Kreisel reviewed Wittgenstein's posthu-

---

[22]We may speculate that when Kreisel said he had been at Dudley, Dyson, not having heard of Dudley Grammar School, took him to have said "Dulwich". Another possibility, which we can't rule out, is that Kreisel told Dyson he had been at Dulwich.

mously published *Remarks on the Foundations of Mathematics*, he expressed disappointment, but at the same time recognized an insight in Wittgenstein's writings that echoed a key idea of his own, namely that "every significant piece of mathematics has a solid mathematical core [...] and if we look honestly we shall see it. That is why Hilbert-Bernays vol.II, and particularly Herbrand's theorem satisfied me: it separates out the combinatorial (quantifier-free) part of a proof (in predicate logic) which is specific to the particular case, from the 'logical' steps at the end. Certain interpretations of arithmetic and analysis have a similar appeal for me." ([Kreisel, 1958a, p.158]).

While an undergraduate Kreisel wrote two short notes for *Eureka*, the Journal of the Archimedeans (The Cambridge University Mathematical Society), "A remark on the Schröder-Bernstein theorem" [Kreisel, 1944a], and "On a geometric trifle" [Kreisel, 1944b], which were published in 1944. These were Kreisel's first publications in mathematics. [I owe reference to these papers to Kenneth Derus.]

During the Second World War, Cambridge undergraduate degree courses could be reduced from the normally required three years of residence, i.e. nine terms, to five terms, on the basis of four terms of residence being 'allowed' for war service[23]. Kreisel completed his examination requirements for the B.A. in Mathematics in five terms by taking Part II in Easter Term 1943, but then stayed on in Cambridge for the following term, i.e. Michaelmas 1943 [message from Jacqueline Cox 1 July 2016]. In accordance with the decree allowing four terms of war service to count as residence, Kreisel was allowed Lent, Easter and Michaelmas 1944 and Lent 1945 (*ibid.*). However, he only needed the three terms in 1944 to fulfill his residence requirement, and so completed his B.A. at the end of Michaelmas Term 1944. Kreisel's Vita in ([Odifreddi, 1996, p.xiii]) accordingly lists his Cambridge B.A. as having been earned in 1944. (He formally took his B.A. on 3 August 1945 by proxy (i.e. *in absentia*) (*ibid.*).)

Kreisel's Jewish Refugees Committee Notes [8] for June 1943 state, "Passed Part 2 of the Tripos with class 1. Senior Wrangler" (case record, Committee of Refugees). This reference to Senior Wrangler, i.e. the top first in Mathematics, is problematic, but cannot be dismissed out of hand, since the person who wrote this, Greta Burkhill, was the wife of Charles Burkill, a Cambridge University Lecturer in Mathematics, so the basis on which she wrote this might have come from him. However, formal designation of a Se-

---

[23]"The Council of the Senate gives notice that they will allow four terms, towards degrees usually requiring nine terms' residence, to students absent during four terms for approved reasons connected with the war" ([2], p.491).

nior Wrangler had been abolished in 1909, when G.H. Hardy, who considered the emphasis on training students to solve tricky mathematical problems during examinations to be a pernicious education for mathematicians, managed to put an end to formal declaration of the top first. Nonetheless, some informal recognition carried on. Verena Huber-Dyson wrote that "Apparently Freeman Dyson, Georg Kreisel, James Lighthill and John Myhill were the major contenders for the Senior Wranglership" ([Huber-Dyson, 1996, p.54]). I wrote to Freeman Dyson asking about this assessment. His answer was, "nobody was in contention. [...] So far as I remember, the question who was Senior Wrangler never arose. That was a historical relic that we had long outgrown." (email message 24 April 2016). As noted above (p.96), Dudley Grammar School had been informed that Kreisel was "placed very high in the list of Wranglers of his year", and that "as a result of his performance in the Mathematical Tripos, Part II, he had been awarded a Senior Scholarship for post-graduate work." At the same time, the accomplishments of Freeman Dyson and James Lighthill in those 1943 exams were extraordinary, having sat and received a Distinction in Part III at the same time as sitting Part II. Kreisel did not sit Part III, but his accomplishment having mastered proof theory to research level in the form of Hilbert-Bernays Volume II (not taught or examined in Cambridge) while studying for the Tripos Part II, along with substantial discussions with Wittgenstein, can be seen as comparable.

4   Researcher in the British Admiralty 1943–1946

Kreisel used his mathematical training in fluid dynamics for war service as an Experimental Officer in the British Admiralty. He entered the service on 18 December 1943 ([13]), and was assigned to a research centre in a converted country house called West Leigh near the town of Havant, not far from the major naval base at Portsmouth ([Crick, 1996, p.25]). Francis Crick, at that time a physicist, was also doing research for the Admiralty at West Leigh, and they became close friends, a friendship that endured to the end of Crick's life, in 2004. Throughout the many years of their friendship, they always addressed each other in correspondence as 'Dear Crick–Dear Kreisel' ([Crick, 1996, p.25]), in the British way of friendly formality. Almost no one ever addressed Kreisel as "Georg".

Matt Ridley [2006] explains that (p.17) "Three people entered Francis Crick's life towards the end of the war. All of them would remain part of his life till the end, and all of them would nudge him toward his future greatness." One of these three was Georg Kreisel, who "was seven years younger than Crick but was to be more mentor than disciple." Crick [1988] turned to Kreisel for advice in deciding what path to follow after his war work in the Admiralty (pp.15–16): "By this time I was reasonably sure that I didn't want to spend the rest of my life designing weapons, but what did I want to do? [...] I was sure in my mind that I wanted to do fundamental research rather than going into applied research, even though my Admiralty experience would have fit me for developmental work. But did I have the necessary ability?" He asked the person under whom he had worked during the war. "I also asked my close friend Georg Kreisel, now a distinguished mathematical logician. [...] By this time I knew Kreisel well, so I felt his advice would be solidly based. He thought for a moment and delivered his judgment: 'I've known a lot of people more stupid than you who've made a success of it.'" (*loc. cit.*). Kreisel's infuence on Crick was ongoing: "I have known him now for about fifty years. Over that time I have been immensely influenced by his powerful intellect" ([Crick, 1996, pp.31–32]). Crick [1996] also acknowledges specific help (p.27)): "He helped me once with a mathematical problem—the Fourier transform of a coiled-coil—on which I was stuck". This problem lay at the heart of the work with Watson on the double-helix (see [Watson, 2012, p.152]).

Crick gives a sketch of Kreisel's war work ([Crick, 1996, p.26]), beginning with the remark "I never did discover exactly what Kreisel was doing at West Leigh. There was a war on and I had plenty of urgent problems of my own to deal with. I believe that one of his first efforts was to apply the methods of Wittgenstein to the problem of mining the Baltic". Later Kreisel was transferred to the Department of Miscellaneous Weapons Development, housed in Fanum House in London, near Leicester Square. Crick reports that there Kreisel "was involved with calculating the effects of waves on the floating (Mulberry) harbors being designed for the Normandy landing", which included supervising some hydrodynamic experiments at Imperial College (*loc. cit*).

There is a first-hand account of Kreisel at the Department of Miscellaneous Weapons Development by Gerald Pawle, a Royal Navy Volunteer Reserve officer in the Department (and a journalist before and after the war—see "The Author", [Pawle, 1956] 2009 edition): "Another notable civilian recruit was George Kreisel, whose mathematical calculations were far above the heads of most members of DMWD. Given a problem, he would retaliate with pages and pages of hieroglyphics which reduced even Richardson to stunned perplexity. And once, when in desperation Richardson asked an eminent professor of mathematics to interpret an explanation of Kreisel's, the learned man confessed with some diffidence that it was altogether too advanced for *him*! Kreisel was certainly a colourful addition to the department. He had his own decided views on the appropriate rig for Service occasions, and would turn up at official trials attired in an old pair of grey flannels and a sky-blue shirt, widely opened at the the neck. This gave him the carefree appearance of a holiday hiker who had somehow strayed into the decorous gatherings of uniformed officers by sheer accident. But George Kreisel himself remained entirely oblivious to the critical stares of the admirals and generals, all of whom seemed to him uncomfortably overdressed" ([Pawle, 1956, pp.191–2]).[24]

The Mulberry harbours for which Kreisel did his calculations at the DMWD and experiments at Imperial College played a significant role in the allied invasion of Normandy, which began on Tuesday, 6 June 1944 (D-Day) (see [20]). They were off-shore harbours used to deliver men and equipment onto the beaches until French ports had been captured and repaired. The components of these harbours were floated across the Channel and assembled *in situ*. Kreisel's calculations and experiments concerned their stability in heavy seas and storms. There were two Mulberry harbours, one at Omaha Beach, Mulberry A, the other at Gold Beach (Arromanches), Mulberry B. According to [20], "Both harbours were almost fully functional when on 19 June a large north-east storm at Force 6 to 8 blew into Normandy and devastated the Mulberry harbour at Omaha Beach. The harbours had been designed with summer weather conditions in mind but this was the worst storm to hit the Normandy coast in 40 years. The destruction at Omaha was so bad that the entire harbour was deemed irreparable. [...] The Mulberry harbour at Arromanches was more protected, and although damaged by the storm, it remained intact." Since this was the worst storm in 40 years, it must have been beyond the force for which Kreisel's calculations were required to demonstrate stability, and on the other hand one can suppose that Kreisel's calculations showed that the harbor at Arromanches was constructed to a standard by which it did weather this storm. Actually to establish what role Kreisel's calculations and experiments on waves played in the survival of Mulberry B requires archival research to find documentation showing that Kreisel's calculations were used in the design of the harbours, or had the effect of causing modifications to be made while their components were being constructed.

"When it was finally completed on D plus 40 [that would be 16 July 1944] the harbour at Arromanches was a truly remarkable enterprise. Two miles long by a mile broad[25], it was maintained by a force of 5000 officers and men of

---

[24]I owe this reference to Richard Hawkins ([Hawkins, 2020, fn.78]).

[25]These dimensions are of the entire harbour, consisting of an outer ring of caisson breakwaters, jetties at which the supply and troop ships docked, and floating roadways from the jetties to the shore; see aerial photographs of Mulberry B in [20].

the Royal Navy, and a fleet of hundreds of specialized craft, including port-construction ships, boom defence vessels, tankers, ferries, floating cranes, and floating docks. The harbour was defended against air attack by nearly 200 Army guns, as well as the guns of the fleet; on the Eastern flank two miles of nets acted as a trap for infernal machines, long-range torpedoes, one-man submarines, and drifting mines" ([Pawle, 1956, pp.286–87]). On 23 July 1944, one week after it was completed, Winston Churchill visited Mulberry B (which came to be called Port Winston) and paid tribute to it: "This miraculous port has played, and will continue to play, a most important part in the liberation of Europe." ([Pawle, 1956, p.287]). "The harbour at Gold Beach was used for 10 months after D-Day and over 2.5 million men, 500,000 vehicles, and 4 million tons of supplies were landed before it was fully decommissioned" ([20]).

Nazi Germany surrendered on 8 May 1945 (VE-Day), and World War II came to an end on 15 August 1945 with the surrender of Japan (VJ-Day). The Department of Miscellaneous Weapons Development was disbanded in the autumn of 1945 ([Pawle, 1956, p.288]), but Kreisel continued to do mathematical work in the Admiralty Research Laboratory until the summer of 1946. He had been a civilian while working in the DMWD, but was now given the Naval rank of Lieutenant (equivalent to Captain in the Army), more precisely (Scientific) Temporary Lieutenant (Special Branch) Royal Navy Volunteer Reserve[26]. Kreisel held this rank from 16 July 1945 ([14]) to 29 April 1946 ([15]). One wonders if he then had to wear a uniform. While at the Admiralty Research Laboratory he wrote a research report "Cavitation with finite cavitation numbers" ([Kreisel, 1946b])[27], and also a report, "Hydrodynamic researches", for the British Intelligence Sub-Committee ([Kreisel, 1946a])[28]. During this first post-war year Kreisel shared a flat with Crick in London (at

---

[26]Richard Hawkins discovered that Kreisel held the rank of Lieutenant from "School Notes" in *The Dudleian*, March 1946 ([5], p.40): "The final report of the Dudley Refugee Committee issued last December closes a chapter of which we may well be proud. [...] Three boys were educated in the school from 1938 to 1941, living at 'Lingwood', then the school boarding-house, under Dr. Mainwaring. [...] Georg Kreisel, after winning an open scholarship in Mathematics to Trinity College, Cambridge, became a 'Wrangler' in the Mathematical Tripos of 1943, and is now a Lieutenant in the Navy". He then managed to find documentation for this fact from relevant issues of *The Navy List*.

[27]I owe this reference to Kenneth Derus, from a supplement to Kreisel's Bibliography he produced.

[28]The existence of this report was discovered by Richard Hawkins. Its catalog entry in the Imperial War Museum Archive doesn't give a date, but Hawkins was able to date it to this period by the fact that its author is given as Lieutenant G. Kreisel, RNVR in [Saunders, 1949, p.87] (email message 15 January 2020).

56 St George's Square, Pimlico) ([Ridley, 2006, p.20]), until he returned to Cambridge at the start of the academic year 1946–47.

5 Return to Cambridge 1946–1949

An entry in Kreisel's Jewish Refugees Committee Notes [8] dated October 1946 reads "Released under class B, returned to Trinity Coll. Cambridge doing Mathematical Research". Class B was one of two routes to demobilization ([Pope, 1995, pp.69–70]). In returning to Trinity, Kreisel took up the Senior Scholarship he had been awarded for his outstanding Part II Tripos results, but he was not enrolled as a graduate student of Cambridge University [email message from Jacqueline Cox 19 April 2016], and was never supervised for a graduate degree. Nonetheless he considered himself a student, later referring to "my student days in Cambridge: before and after my military service (in England)" ([Kreisel, 1989, p.148]).

Back in Cambridge, Kreisel returned to his interests in logic and philosophy, though he also published two substantial papers in hydrodynamics arising from his work for the Admiralty, [Kreisel, 1949a] and [Kreisel, 1949b], both submitted in January 1948, with [Kreisel, 1949a] published in April 1949, and [Kreisel, 1949b] in June 1949. These were to do with surface waves, one with that title, the other with contributions to the mathematics behind that part of hydrodynamics. These are Kreisel's first papers published in a professional mathematical journal. A glance at them shows what Gerald Pawle was talking about when he referred to "pages and pages of hieroglyphics", and said that Kreisel's "mathematical calculations were far above the heads of most members of DMWD".

While Kreisel did not carry on further research in hydrodynamics after those two papers, he took the work he had done in this subject to heart in his philosophical thinking, of which a characteristic example is the section "'Deep' processes" of his paper "On some conversations with Wittgenstein: recollections and reflections" [Kreisel, 1989].

> In the war (at least, after early 1944) I was occupied with hydrodynamics: among other things, with artificial harbors and sea-waves near the coast. One wanted to record the wave motion by measuring the pressure at the bottom of the sea (such measurements being inexpensive). Now, the theory looks like this. The (tacitly: exact) distribution of pressure on the bottom determines (also exactly) the so-called potential, and thereby the different aspects of wave motion. But here the experts actually over-

looked the fact (so it was not just the mere *possibility* of an oversight) that small variations (i.e. errors) in the distribution of pressure can go with large variations in the shape and motion of the sea surface. Pedantically: aspects that strike the eye (e.g. the height of the waves) remain undetermined. When the waves are short (compared to the depth of the sea) the pressure at the bottom of the sea is slight, even if the waves themselves are relatively high. (p.153)

Kreisel makes a direct link between the phenomenon he has just described and a point he brought up in his reading of G.H. Hardy's *A Course of Pure Mathematics* [Hardy, 1921] with Wittgenstein in 1942, on giving constructive content to the intermediate value theorem—which Hardy calls (p.216) the fundamental property of a continuous function. (This comes in [Kreisel, 1989, pp.150–51], in which he reports making the point that very small changes in the function can result in very big changes in the intermediate value. Kreisel goes on to remark that "For me, this story of the waves on the surface and the pressure deep down remains a vivid metaphor, in particular for self-protection against gushing about depth, which to me is a kind of intellectual pollution." (*op. cit.*, p.154).

Kreisel's senior scholarship gave him accommodation in college[29], with a small stipend that was not enough to live on. To support himself, Kreisel "for a brief period [...] taught the boys at an English boarding school" ([Crick, 1996, p.28]), and Crick reports Kreisel having told him that the boys there "referred to him as 'The Pwoblem Child' because he would frequently say in class, 'What is the pwoblem?'" (*loc. cit.*).

Kreisel pursued his research in mathematical logic and resumed discussions with Wittgenstein. He was present at the (in)famous meeting of the Moral Sciences Club on 25 October 1946 when Karl Popper gave a talk with which Wittgenstein disagreed. His recollection of the event emphasized different aspects from those that became notorious [Edmonds and Eidinow, 2001]. On 24 February 1947 Kreisel was naturalized as a citizen of the United Kingdom ([8], supplementary card). On 27 February 1947 Kreisel gave a talk to the Moral Sciences Club (Sixth Meeting in Lent Term) on "Mathematical Logic", chaired by Wittgenstein ([12], p.148)[30]. During that term Georg Henrik von Wright returned to Cambridge (he had previously been in Cambridge during

---

[29] According to the University Residents Lists (as reported in an email message from Jonathan Smith 26 April 2016), he was allocated rooms in Trinity for the academic years 1946–7 and 1947–8.

[30] This fact, as noted in [Klagge and Nordmann, 2003, p.355], was drawn to my attention by Akihiro Kanamori.

the 1938–39 academic year, when he attended Wittgenstein's lectures on the foundations of mathematics), and Kreisel got to know him. Wittgenstein wrote to von Wright 24 May 1947: "I was very glad to read the kind words you said in your letter about Kreisel and Miss Anscombe. (I, too, respect both of them very much.)" ([Wittgenstein, 1983, p.57, and p.63 for the date]).

Kreisel remained on friendly terms with von Wright, and in November 1957 sent him a draft of his review of the publication of Wittgenstein's *Remarks on the Foundations of Mathematics* [Kreisel, 1958a]. Von Wright responded constructively but critically, and Kreisel made revisions in response, though strong disagreements remained between them about how to understand Wittgenstein's philosophy of mathematics and his philosophy more broadly[31]. Kreisel contributed to a Festschrift for von Wright in 1976 [Kreisel, 1976].

In July 1947 Kreisel wrote to Paul Bernays, the beginning of a correspondence which continued without break until June 1977, shortly before Bernays died (see the catalogue of Bernays correspondence with Kreisel in [1]). In his first letter to Bernays (from Trinity College Cambridge), dated 3 July 1947, Kreisel tells Bernays that he has written a paper "The Skolem model and undecidable propositions in the calculus of predicates"—evidently an early version of his first paper in mathematical logic that would be published in 1950 as "Note on arithmetic models for consistent formulae of the predicate calculus" [Kreisel, 1950]—which makes use of the section in Hilbert & Bernays volume 2 [Hilbert and Bernays, 1939, pp.234–253], in which Bernays applies the method of arithmetization to the Gödel Completeness Theorem. Kreisel gives a brief description of his results, and asks permission to send details, which he does in an eight page typed letter dated 2 December 1947.

Kreisel's relationship with Bernays was such that when Kreisel considered visiting the Institute for Advanced Studies for contact with Gödel, he asked Bernays to send a letter of reference, and on 27 January 1955 Bernays wrote to Freeman Dyson (from 1953 a permanent member of the Institute for Advanced Studies), supporting Kreisel's successful application to be a visitor at the IAS. In 1958 Kreisel contributed to a Festschrift for Bernays on the occasion of his 70th birthday, with a paper on Hilbert's programme [Kreisel, 1958b], in a special issue of the journal *Dialectica* (which lent its name to the interpretation of arithmetic using primitive recursive functionals of finite type from Gödel's contribution to that volume)

Wittgenstein resigned his Cambridge professorship at the end of the academic year 1946–47 and was succeeded by Georg Henrik von Wright. Iris

---

[31] I am grateful to Andrew English for drawing this correspondence to my attention.

Murdoch arrived in Cambridge as a postgraduate philosophy student at Newnham College for the academic year 1947–48. She had hoped to study with Wittgenstein, but had to content herself with supervision from John Wisdom and getting to know those who knew Wittgenstein, among them Kreisel ([Conradi, 2001, p.262]). Iris Murdoch and Kreisel became important friends to each other in that year, a friendship which endured until she began to succumb to Alzheimer's disease in 1996. Iris Murdoch dedicated one of her novels to Kreisel (*An Accidental Man*, published in 1971), and a number of characters in her novels are said to be based on aspects of Kreisel's personality ([Conradi, 2001, p.265]).

Francis Crick, transitioning from physics to biochemistry, left the Admiralty in September 1947 and moved to Cambridge ([Ridley, 2006, p.29]), where he and Kreisel saw a good deal of each other[32]. Crick [1996] recounts the following story (p.29): "One day I went into my office at the Cavendish and was surprised to see a letter for me with a Spanish postmark. Inside was a picture of Kreisel, beaming away between two rather good-looking young men. I do not read Spanish but I could understand enough to tell [...] that the letter, although addressed to me by name, was really for Kreisel. When I saw him next I tackled him about it. He was quite unabashed. 'When I travel,' he said, 'I often use your name.' Being good friends I was only amused by this little habit of his." Matt Ridley adds to this story that "On another occasion, when arrested on a Moroccan beach, Kreisel gave the police Crick's name" ([Ridley, 2006, p.18]).

Kreisel submitted a dissertation for a Title A (Research) Fellowship in Trinity College for election from October 1947 (one of 24 candidates, of whom 7 were successful) [email message from Jonathan Smith 13 April 2016], but was not elected. Trinity retains no records of unsuccessful applications, so we don't know the title of the dissertation he submitted. It might have been an early version of his first published paper in mathematical logic, "Note on arithmetic models for consistent formulae of the predicate calculus" [Kreisel, 1950], which initiated a deeper understanding of the incompleteness phenomenon first established by Gödel (see Walter Dean [2019] for an excellent account of what is accomplished by Kreisel in this paper, and its context). However, it might instead have been an early version of his second published paper in mathematical logic, "On the interpretation of non-finitist proofs–Part I" [Kreisel, 1951]. This paper initiated what came to be called Kreisel's Programme, the aim of which was to extract finitist content from non-finitist

---

[32]The last entry in Kreisel's Jewish Refugees Committee Notes is dated August 1947 and reports him as "Remaining at Trinity College for research."

proofs (also later called proof unwinding), which Kreisel began working on after he returned to Cambridge in 1946. As an undergraduate he had (see above p.101) been "attracted to the logical wit of consistency proofs" from reading Hilbert-Bernays Vol.2. By the time he returned to the study of mathematical logic after the war, he considered that it was dubious to doubt the consistency of a well understood mathematical theory (a harbinger of his later thinking in terms of informal rigour—see [Kreisel, 1967] and [Kreisel, 1987b]), but rather than dismissing Hilbert's programme, Kreisel saw a way to repurpose it as a means of extracting finitist content from non-finitist mathematical proofs: "After the war I had a chance to go into mathematical logic in more detail; in particular, into consistency proofs. Instead of pursuing Hilbert's aim of eliminating dubious doubts about the usual methods of mathematics, a more compelling application (better: interpretation) of those proofs occurred to me." ([Kreisel, 1989, p.151]). The idea that there was such finitist content to be extracted, Kreisel attributed to Wittgenstein (see p.102 above), though of course not the idea of how to find it by means of Hilbertian proof theory.

6 Appointment at Reading University and first publication in mathematical logic 1949–1950

In 1948 Kreisel applied for and was appointed to a Lectureship in Mathematics at Reading University, starting in 1949. (This appointment was in succession to R.L. Goodstein, another Cambridge mathematics undergraduate with strong links to Wittgenstein. Goodstein left Cambridge in 1935 to take up his appointment at Reading, a position he held until December 1947, when he left for a chair at the University of Leicester ([Rose, 1988, p.159]).) Crick recalls talking to Kreisel after he returned from his interview for the Reading job. " 'How did it go?' I asked. After a pause he replied, 'By the end they were thinking before they asked a question.' " Crick [1996] remarks that (p.28) "In spite of this they elected him and he moved to a suburban house in Reading." Kreisel's first published paper in logic [Kreisel, 1950] appeared the following year, and listed the author as G. Kreisel (Reading, England).

Acknowledgements

I am deeply indebted in writing this paper to Kenneth Derus for his generosity in sharing the cache of material about Georg and Fritz Kreisel he obtained from the World Jewish Relief's Archives, including [8] and [9] and the brothers' entrance cards to the U.K. (Georg Kreisel's is reproduced on p.89), for sharing information about what happened to Kreisel's parents including [10], for his continual readiness to discuss issues that arose while writing this paper, for helping to find material I was looking for, and for reading and correcting successive drafts. I am also highly indebted to Richard Hawkins for generously sharing invaluable documents ([7], [11], [5], [13], [14], [15]) he obtained in his research on the Dudley Refugee Committee and the Kindertransport, for sending me a copy of his paper [Hawkins, 2020] in advance of publication, and for generously sharing valuable information in response to questions I asked in correspondence. I am tremendously indebted to Freeman Dyson and extremely grateful for his ready willingness to enter into correspondence about Kreisel, for copying to me his correspondence with Kreisel, and for his kind permission to use the material I quote from him in this paper. I am greatly indebted to Elaine Nicholson, Assistant Archivist at the Dudley Archives, who found for me various Dudley Grammar School records and all the material from *The Dudleian* I've made use of in this paper, apart from [11] and [5]. I am very grateful to Hans-Peter Leeb for his unstinting work in editing this paper. I owe a tremendous debt of gratitude to my wife, Kassandra Isaacson, for her huge support to me in the writing this paper, for which, as always, I am hugely grateful. Others to whom I am indebted and grateful for their help are Marianna Antonutti Marfori, Matthias Baaz, Alan Buckley, Jacqueline Cox (Deputy Keeper of Archives Cambridge University Library), Andrew English, Peter Hacker, Akihiro Kanamori, Sally Kent (Assistant Archivist Cambridge University Library), Angus Macintyre, Paolo Mancosu, Brian McGuinness, Dana Scott, Jonathan Smith (Archivist Trinity College Cambridge), Robert Thomas, Robert Tragesser, Mark van Atten, and Alex Wilkie.

References

Sources with known authors

[Alcorn, 1999] Noeline Alcorn. *To the Fullest Extent of his Powers: C.E. Beeby's Life in Education*. Victoria University Press, Wellington, 1999.

[Clapham, 1943] Sir John Clapham, Chairman of the Cambridge Refugee Committee. Memorandum submitted to the General Committee for Refugees on Education of Young Refugees Above the Age of 16. 18 November 1943.

[Conradi, 2001] Peter Conradi. *Iris Murdoch: A Life*. Harper-Collins, London, 2001.

[Crick, 1988] Francis Crick. *What Mad Pursuit: A personal View of Scientific Discovery*. Basic Books, New York, 1988.

[Crick, 1996] Francis Crick. Georg Kreisel: a Few Personal Recollections. In [Odifreddi, 1996], pages 25–32.

[Dean, 2019] Walter Dean. Incompleteness via Paradox and Completeness. *Review of Symbolic Logic*, 1–52, published online 23 May 2019.

[Huber-Dyson, 1996] Verena Huber-Dyson. Thoughts on the Occasion of Kreisel's 70th Birthday. In [Odifreddi, 1996], pages 51–73.
[Dyson, 2018] Freeman Dyson. *Maker of Patterns: an Autobiography Through Letters*. Liveright Publishing Corporation, 2018.
[Edmonds and Eidinow, 2001] David Edmonds and John Eidinow. *Wittgenstein's Poker: The story of a ten-minute argument between two great philosophers*. Faber and Faber, London, 2001.
[Hardy, 1921] G.H. Hardy. *A Course of Pure Mathematics*. Cambridge University Press, 3rd edition, 1921.
[Hawkins, 2020] Richard A. Hawkins. The Dudley Refugee Committee and the Kindertransport, 1938–1945. In *Jewish Historical Studies: Transactions of the Jewish Historical Society of England*, 51:183–201, 2020.
[Heitler, 1961] Walter Heinrich Heitler. Erwin Schrödinger: 1887–1961. *Biographical Memoirs of Fellows of the Royal Society*, 7:221–226, 1961.
[Hilbert and Bernays, 1939] David Hilbert and Paul Bernays. *Grundlagen der Mathematik Zweiter Band*. Julius Springer Verlag, Berlin, 1939.
[Klagge and Nordmann, 2003] James Klagge and Alfred Nordmann, editors. *Ludwig Wittgenstein: Public and Private Occasions*. Rowman & Littlefield Publishers, 2003.
[F. Kreisel, 1940] Fritz Kreisel. The Swing Beam. *The Dudleian*, issue 106:79–80, March 1940.
[Kreisel, 1939] Georg Kreisel. The Tenth of November: a Recollection. *The Dudleian*, issue 104:101–103, July 1939. Reprinted as an Appendix to this paper, pp.117–119.
[Kreisel, 1941] Georg Kreisel. The Devil. *The Dudleian*, issue 110:123–124, July 1941.
[Kreisel, 1944a] Georg Kreisel. A remark on the Schröder-Bernstein theorem. *Eureka: The Journal of the Archimedeans (The Cambridge University Mathematical Society)*, *Eureka*, 8:8-9,1944.
[Kreisel, 1944b] Georg Kreisel. On a geometric trifle. *Eureka*, 8:8–9, 1944.
[Kreisel, 1946a] Georg Kreisel. Hydrodynamic researches. *British Intelligence Objectives Sub-Committee, 1945–46*, Imperial War Museum Catalogue number LBY K. 27023-7.
[Kreisel, 1946b] Georg Kreisel. Cavitation with finite cavitation numbers. *British Admiralty Research Laboratory*, Report No. R1/H/36, 1946.
[Kreisel, 1949a] Georg Kreisel. Surface waves. *Quarterly of Applied Mathematics*, 7(1):21–44, April 1949.
[Kreisel, 1949b] Georg Kreisel. Some remarks on integral equations with kernels: $L(\xi_1 - x_1, \ldots \xi_n - x_n; \alpha)$. *Proceedings of the Royal Society A*, Volume 197 (June 1949), pp.160–183.
[Kreisel, 1950] Georg Kreisel. Note on arithmetic models for consistent formulae of the predicate calculus. *Fundamenta Mathematicae*, 37:241–267, 1950.
[Kreisel, 1951] Georg Kreisel. On the interpretation of non-finitist proofs—Part I. *The Journal of Symbolic Logic*, 16:241–267, 1951.
[Kreisel, 1958a] Georg Kreisel. Review of Wittgenstein's *Remarks on the Foundations of Mathematics*. *The British Journal for the Philosophy of Science*, 9:135–158, 1958.
[Kreisel, 1958b] Georg Kreisel. Hilbert's programme. *Dialectica*, 12:142–168, 1958. Reprinted with revisions, added notes, and a Postscript (Autumn 1978) in Paul Benacerraf and Hilary Putnam, editors, *Philosophy of Mathematics: Selected Readings*, pages 207–238. Cambridge University Press, second edition, 1983.

[Kreisel, 1960] Georg Kreisel. Wittgenstein's theory and practice of philosophy (review of *The Blue and Brown Books*). *The British Journal for the Philosophy of Science*, 11:238–252, 1960.

[Kreisel, 1967] Georg Kreisel. Informal rigour and completeness proofs. In Imre Lakatos, editor. *Problems in the Philosophy of Mathematics: Proceedings of the International Colloquium in the Philosophy of Science, London, 1965, volume 1*, pages 138–171. North-Holland Publishing Co., Amsterdam, 1967.

[Kreisel, 1976] Georg Kreisel. Der unheilvolle Einbruch der Logik in die Mathematik. *Acta Philosophica Fennica (Essays on Wittgenstein in Honour of G.H. Von Wright)*, 28:166–187, 1976.

[Kreisel, 1978] Georg Kreisel. Zu Wittgensteins Gesprächen und Vorlesungen über die Grundlagen der Mathematik. In Elisabeth Leinfellner et al., editors. *Wittgenstein and His Impact on Contemporary Thought: Proceedings of the Second International Wittgenstein Symposium 29 August to 4 September 1977*, pages 79–81. Hölder-Pichler-Tempsky, Vienna, 1978. Partial English translation in [Klagge and Nordmann, 2003, p.355].

[Kreisel, 1987a] Georg Kreisel. Proof theory: some personal recollections. In Gaisi Takeuti, editor. *Proof Theory*, pages 395–405. North-Holland Elsevier, Amsterdam, second edition, 1987.

[Kreisel, 1987b] Georg Kreisel. Church's thesis and the ideal of informal rigour. *Notre Dame Journal of Formal Logic*, 28:499–519, 1987.

[Kreisel, 1989] Georg Kreisel. Zu einigen Gesprächen mit Wittgenstein: Erinnerungen und Gedanken. In Michael Huter, editor. *Wittgenstein: Biographie-Philosophie-Praxis; Eine Austellung der Wiener Secession*, pages 131–143. Wiener Secession, Vienna, 1989. English translation by W.B. Ewald with emendations by Kreisel, "On some conversations with Wittgenstein: recollections and reflections", in [Odifreddi, 1990, pp.148–156].

[Kreisel, 1998] Georg Kreisel. Second thoughts around some of Gödel's writings: A non-academic option. *Synthese*, 114:99–160, 1998.

[Mainwaring, 1940] James Mainwaring. *Man and His World: A Course in History and Geography (with illustrations in black and white by the author and others)*, Book I *The Evolution of the Old World*, George Philip & Son, London, 1940; Book II *The Evolution of the Modern World*; Book III *The World and Its Wealth*, George Philip & Son, London, 1949; there were subsequent editions up to a 4th edition in 1960.

[McGuinness, 2012] Brian McGuinness, editor. *Wittgenstein in Cambridge: Letters and Documents 1911–1951*. Wiley-Blackwell, 2012.

[Monk, 1990] Ray Monk. *Ludwig Wittgenstein: The Duty of Genius*. Jonathan Cape, London, 1990.

[Horner and Rowe, 2015] Avril Horner and Anne Rowe, editors. *Living on Paper: Letters from Iris Murdoch 1934–1995*. Chatto & Windus, London, 2015.

[Odifreddi, 1990] Piergiorgio Odifreddi, editor. *About Logic and Logicians: A Palimpsest of Essays by Georg Kreisel, Volume 1*. Unpublished, 1990. Published online by Rodrigo Freire, Lógica no Avião, Brasília, 2019.

[Odifreddi, 1996] Piergiorgio Odifreddi, editor. *Kreiseliana: About and Around Georg Kreisel*. AK Peters, Wellesley, Massachusetts, 1996.

[Pawle, 1956] Gerald Pawle. *The Secret War 1939–1945*. George G. Harrop, London, 1956. Reprinted as *The Wheezers & Dodgers: The Inside Story of Clandestine Weapons Research in World War II*, Seaforth Publishing, Barnsley, 2009. Online at https://archive.org/details/secretwar193945007234mbp

Georg Kreisel: Some Biographical Facts 115

[Pevsner, 1974] Nikolaus Pevsner. *The Buildings of England: Staffordshire*. Penguin Books, Macmillan, London, 1974.
[Pope, 1995] Rex Pope. British Demobilization after the Second World War. *Journal of Contemporary History*, 30:65–81, 1995.
[Raybould, 2010] Trevor Raybould. *Dudley Grammar School 1562–1975: A History of the School in its Times*. Bassett Press, Southampton, 2010.
[Ridley, 2006] Matt Ridley. *Francis Crick: Discoverer of the Genetic Code*. Harper Press, London, 2006.
[Rose, 1988] H.E. Rose. Obituary R.L. Goodstein. *Bulletin of the London Mathematical Society*, 20:159–166, 1988.
[Saunders, 1949] Harold Eugene Saunders. *Report of an Inspection of Certain European Hydromechanical and Aeromechanical Test and Research Facilities During October-December 1945, Volume 1*. Washington D.C., 1949.
[Temple, 1962] David Crighton Temple. *Dudley Grammar School: A Chronicle of Four Centuries (1562–1962)*. Published by Dudley Grammar School, 1962. A revised and enlarged edition of [Watson, 1926].
[Watson, 1926] Hugh Watson. *A Short History of Dudley Grammar School, 1562–1926*. Published by Dudley Grammar School, 1926.
[Watson, 2012] James Watson. *The Double Helix*. Annotated and illustrated edition, Simon & Schuster, New York, 2012.
[Wittgenstein, 1953] Ludwig Wittgenstein. *Philosophical Investigations*. Basil Blackwell, Oxford, 1953.
[Wittgenstein, 1958] Ludwig Wittgenstein. *The Blue and Brown Books*. Basil Blackwell, 1958.
[Wittgenstein, 1983] Ludwig Wittgenstein. Some Hitherto Unpublished Letters from Ludwig Wittgenstein to Georg Henrik von Wright. *The Cambridge Review*, Volume 104, Number 2273:56–64, 28 February 1983.

Sources without known authors

[1] Paul Bernays Verzeichnis: Manuskripte – Fremdmanuskripte – Korrepsondenz – Biographisches, ETH Zurich.
    `https://doi.org/10.3929/ethz-a-000381167`
[2] *Cambridge Reporter*, 16 January 1940.
[3] "Conscription: the First World War", *www.parliament.uk: Living Heritage*
    `https://www.parliament.uk/about/living-heritage/`
    `transformingsociety/private-lives/yourcountry/`
    `overview/conscription/`
[4] Dachau Concentration Camp Records available online from
    `https://www.jewishgen.org/databases/holocaust/`
    `0050_DachauIndexing.html`
[5] "School Notes", *The Dudleian* Vol.43 No.2 (March 1946), p.40
[6] Google Street View of Dudley Grammar School Buildings
    `https://www.google.co.uk/maps/@52.5124637,-2.0905628,3a,75y,9.55h,`
    `89.86t/data=!3m6!1e1!3m4!1spxBL4JD67jeM8JhlxU_2xA!`
    `2e0!7i13312!8i6656?hl=en`
[7] *Adressenbuch der Landeshauptstadt Graz und der angrenzenden Gemeinden*, Verlag "Styria", Graz, 1938.
[8] Jewish Refugees Committee Notes (No.3014) on Georg Kreisel.
[9] Jewish Refugees Committee Notes (No.3015) on Fritz Kreisel.

[10] Probate of Bertha Kreisel's will, *National Probate Calendar for England and Wales (Index of Wills and Administrations), 1858–1995*, 1951, p.189.
[11] tribute to Dr. James Mainwaring on his departure from Dudley Grammar School December 1944, *The Dudleian* Volume 42 No.1 (December 1944), pp.11–12.
[12] Minute book of the Moral Sciences Club 1935–1952, Cambridge University Library, UA Min.IX.44.
[13] *The Navy List July 1945*, p.2307.
[14] *The Navy List October 1945*, p.587.
[15] *The Navy List July 1946*, p.2459.
[16] Office of National Statistics Average UK House Prices
https://assets.publishing.service.gov.uk/government/uploads/system/uploads/attachment_data/file/305683/Table_502_-_ONS.xls
[17] "Snails are making a comeback in Austria"
https://www.saveur.com/austrias-culinary-snail-scene-is-making-comeback/
[18] St James Academy, Dudley website
https://www.stjamesacademy.org.uk/page/?title=About&pid=6
[19] Wikipedia, "County Borough of Dudley",
https://en.wikipedia.org/wiki/County_Borough_of_Dudley
[20] Wikipedia, "Mulberry harbours",
https://en.wikipedia.org/wiki/Mulberry_harbour
[21] Wikipedia, "Wrangler (University of Cambridge)",
https://en.wikipedia.org/wiki/Wrangler_(University_of_Cambridge)

Appendix

from *The Dudleian* issue 104 (July 1939), pp.101–103

THE TENTH OF NOVEMBER

A RECOLLECTION

It was the result of six years of National Socialist rule. Press and Party have done their best to convince people of the responsibility of the Jews for all difficulties Germany has now to face: the Treaty of Versailles, the lack of raw materials, the national debts and the inflation, it has consistently been suggested were caused by "the Jews." A propaganda ministry, for which the budget provides almost as much as for the armaments, has been formed and all means have been tried to spread this view. It has been promised that all the possessions and offices formerly held by Non-Aryans shall be handed over to unemployed party members. The racial theory has been approved by the government and the nation is taught that, if a Jew happens to do anything wrong, not the individual but the whole race is responsible. (By the way, even the definition of the term "Jew" is still a matter of controversy). In Germany Jews have been deprived of all civil and even social rights only because of their Semitic origin; on the other hand, newspapers show warm sympathy for the poor Arabs in Palestine who are murdered by the "cruel Britons." Of course, party leaders emphasize again and again that no criticism or even discussion of the Nazi doctrine is desired.

This eager preparation of public opinion has fogged the spirit of the masses. The youths have been dressed in uniforms; from morning till evening they are under the control of the party—first at school and then in drill; practically no foreign newspapers are available and the German press has been assuring the population of the continuous and considerable progress of Herr Hitler's policy. Shortly after he had obtained power it was forbidden to listen to most foreign wireless stations and in this way public opinion has been modelled.

Then Hitler's experiment with Sudetenland stimulated all his followers. Long diplomatic negotiations delayed the invasion; there was nothing which resembled the glorious march into Austria when sunshine and cheering greeted the German troops. Much money was wasted in propaganda among people in Czechoslovakia and in building the Siegfried Line on the West Front; the budget was in disorder. Now Herman Goering remembered the Vermoegensangabe (complete property exposure) [declaration of assets] of the Jews in the whole Reich and saw an opportunity. By chance some criminal madman assassinated an official in the German embassy in Paris and revenge was justified. Besides, people's attention was led to a horrible pogrom, astonishing in the twentieth century. And it is still the Jewish question which constitutes a refuge for the Nazi government, if any political action fails.

Step by step this poisoning of the German people was carried out and now the glorious work could start! I myself had a glance at the events of this day and I should like to use my own experiences to illustrate a chapter of Germany's most recent history.

Already by the change-over in March 1938 I had lost all my prospects for [a] future in Germany and realized that the lot of emigration, an enforced exile from my home, had become inevitable for me at last. Excluded from public life, I passed the following months like a homeless wanderer, not knowing what [the] future had in store for me. But I was wrong when thinking that thereby my share in the National Socialist regime was ended. Having been unable to attend a school at Graz I went to Vienna to make myself forget all these troubles by work, and entered the only secondary school for Non-Aryans in former Austria.

One day, my turn came. It was a dull day in November. Clouds hid the sun, the town was dressed in grey. I had come to school as usual, and started work, when gradually a certain feeling of anxiety alarmed boys and masters. Parents came and boys left school; from far and near detonations reached the ear. Mother, who happened to be in Vienna then, came to fetch me too; I had already passed the gates of the school, when two men watching for boys before the school building placed themselves before me, and, without saying a word, dragged me to the next "SA Kaserne" (Stormtroopers' barracks). Each of the captives, of whom a great number had already been there before, was sent into a darkened room.

Terrified men awaited with dreadful anticipation the coming hours. Here I first saw with my own eyes the inhuman cruelty which spared neither aged people nor youths. We were driven into a small, stuffy room where we had to remain pressed like herrings.

In this condition we saw before ourselves a series of violences. We were actually loaded on to large lorries standing before the SA Kaserne where a great crowd had gathered. Many a look of pity was cast at us as consolation, but there were many who gazed bloodthirstily: nor were they disappointed.

Now the SA handed in their booty to the SS (Black Guards) for further dealing. These men are specially trained to free the régime of its enemies by any possible means. You know that in Germany in the end of the fifteenth century "Landsknechte" (lansquenets) constituted the Empire's army. Their rough laws and customs have been renewed by Herr Himmler's youths: running the gauntlet brought us more less into a condition of stoic indifference. As for myself, conscious only of having to run through the two formidable rows of SS men, I was carried away by the stream of prisoners, while Black Guards struck to the ground with their helmets all those whom they blindly hit. This brutal beating and slaying did not take place once only, but several times.

Inside the SS Kaserne which one leaves only for the concentration camps, there was already present the atmosphere of the near future. Dark and close was the room, where thousands of people spent hour after hour unable to move a foot's space. Then in the cold, somber night all were driven into the courtyard and it was only shadows who heard the continuous threats, which pierced the dead of the night. At this time, when my mind was numb and I had ceased to care, I was luckily dismissed, together with two other boys—three rescued from 3000 men! Behind me the gates of a hell closed, and, horror-stricken, I sought my way home. The others had the worst before them. Vans with human loads rolled from all parts of the Reich to Dachau, Oranienburg and Sachsenhausen.

Any account of this unforgettable event might lead, I am sure, to a wrong point of view of the present situation of the German people. I think that even by hardest oppression and terror a whole nation cannot be altered. So, while the government had expected a growth of popularity among people after this event, the contrary was achieved, as was inevitable.

G. KREISEL, VUA

[The abbreviation 'VUA' stands for 'Form V Upper Arts'—see "School Notes", *The Dudleian*, issue 104 (July 1939), which records Kreisel's arrival at Dudley Grammar School on April 1st, 1939 with this status.]

6
Chitchat with the Devil: Kreisel's Letters, 2002–2015
KENNETH DERUS

>I'd say there was the *literary idea* of *chitchat with the devil*, where there are no holds barred.
>
>Kreisel to Derus, 14 October 2003

>I know what happiness you have brought him in the last few years (he has occasionally sent me, with your implied permission, your stimulating comments on his memoranda—literary, philosophical, scientific, mathematical etc.) and he says you have an uncanny insight into his ways of thought, which is more than I can claim!
>
>Chancellor to Derus, 26 November 2007

>Kreisel appeared wonderfully serene and happy in a charming period apartment belonging to his fellow mathematician, Baaz. This serenity and happiness is due unequivocally, I am convinced, to his relationship with you and Macintyre. He loved the ritual of receiving your emails from and giving his MSS to a very nice mathematical student, who seemed to call daily. Similarly, he delighted in Macintyre's telephone calls. My own function in our sixty year friendship has been to provide a bit of light relief from his musings on mathematics and logic and related subjects!
>
>Chancellor to Derus, 8 October 2008

Kreisel wrote me 2,834 letters during a 13-year period that ended a month before his death. On top of that, he sent me copies of 1,215 of his letters to other people.[1]

I've dished up extracts from the letters that showcase—not only Kreisel's shop talk, but also—such things as intimacy, anecdote, wit, mastery, literary elegance, autobiography, *caractères* (à la Bruyère), and lessons worth learning.

"On balance I feel the human touch wastes time if one has something interesting to say (of course, almost as a corollary, it can be a Godsend if one

---

[1] The typed correspondence—including my letters to Kreisel and editions of 44 of his essays—weighs in at 5,680 pages (in 36 volumes copyrighted jointly). It and 1,002 pages of unpublished third-party material exist in searchable form.

doesn't")."[2] Kreisel has plenty of interesting things to say, but I've emphasized his human touch because it's what matters most to me.

Readers in Kreisel's line of work will know what and who he's talking about, and others will find answers on the internet. Footnotes labeled *GK* are Kreisel's.

28 May 2002 to Kenneth Derus. I do *not* remember having met CT (Clifford Truesdell) though I have met TC (Truman Capote). Since people tell all sorts of stories—not especially, but also—about me it would be thoughtless (of me) to speculate about your informant. But if not personal acquaintance is meant, I not only KNOW something of CT, I have mentioned several passages in his An idiot's fugitive essays on science, to several people (also in correspondence) especially in the last decade. Let me mention a couple since this is little trouble (to me), and might possibly be of use (to you).

4 February 2003 to Kenneth Derus. Before Horace wrote *Omne tulit punctum qui miscuit utile dulci* he had written *Aut prodesse volunt aut delectare poetae* (with the exclusive AUT). Goethe changed AUT to ET, albeit as a motto for marionettes (or some other dolls), where entertainment is the only use (in erudite jargon aka utility).

I wonder if I never mentioned that I am tone deaf. I discovered it at the age of 12, and realized that I can't expect to hear most of what strikes the musical ear. By a very mild extension, certainly not out of the range of a normal teenager, this realization prepared me for the fact that, depending—not only on education, but also—on all sorts of resources, people will see different aspects of an object in view.

2 July 2003 to Hubert Faure. Many have said of Weil what was recently written of Crick: *Offending someone was always preferable to avoiding the truth*. It may be unusually self-centered (and not fitting the quip about the mote and the beam) to remark that nobody has ever said this about me to me, but many agreed (when asked) that a better quip in my case would be (Quintilian's): Potius amicum quam dictum perdendi.

10 July 2003 to Alex Wilkie. Gödel told me that the degree of my LACK of constant feelings was exceptional. *Remark* He had tried to argue that he was always consistent, and I asked him if he was not disappointed by having learnt

---

[2]Kreisel to Macintyre, 7 February 2006

nothing since his teens. On reflection he recognized that this was a point of view, but not his; a few days later he told me of my fickleness.

15 July 2003 to Kenneth Derus. *Anecdote* around Tarski who insisted on every idea of his being acknowledged. Gandy told the story that, after a short lecture I gave at the ICM 1954 Tarski complained of such a lapse by me; in fact, I had no idea of the publication he meant. Instead I apparently gave another reason: 'There is no need to quote you because everybody knows how clever you are.' According to Gandy, Tarski was pleased. [...]

Your paragraph about *On the idea(l) of logical closure*[3] does not surprise me (and, being in a reasonably good mood), does not even trouble me by CORRECTLY drawing attention to its anticipating my recent letters—not only—to you. But perhaps you'd be surprised by all the gushing drivel I have heard about this article (including its 'biographical' interest) without any specifics like yours, which happen to fit my feelings about it.

21 July 2003 to Kenneth Derus. *Main part* relating—what Goethe considered—his particular gift (for felicitous words without suitable knowledge, but possibly with both thoughts and feeling) to my quip about Dummett,[4] which you seem to have liked for a considerable time: showing that—at least in this respect—you are not fickle (in contrast to me). My quip was (I am pretty sure) high-spirited, if not as elegant as the high spirits of a foal (but not yet as depressing as those of an old cart horse). However, I at least can't help remembering here the famous quip in Goethe's conversation with Eckermann (6.V.1827) about *Faust*, when Eckermann asked Goethe about *an* abstract guiding idea in *Faust*, and Goethe answered to the effect: How should I know? NB Admittedly, in Dummett's book it was not an abstract idea, but a specific theorem that he had missed.

25 July 2003 to Kenneth Derus. Many thanks for your letter of 15.VII., which crossed mine of the same date. The details you give suit me very well; both when they confirm my impressions (especially an understanding, when, by experience,[5] I am never confident), and, of course, when they are new to

---

[3] See [Kreisel, 1992].

[4] See [Kreisel, 1984]: "[...] M. Dummett stumbled on the completeness of Heyting's rules for the positive fragment, but barely noticed this in the maelstrom of his prose."

[5] People innocently tell me they are surprised by what I say (to them) in connection with their own remarks; innocent inasmuch as they don't seem to realize the consequence: *their* understanding of their own remarks is different from mine. – Confirmation brings me noth-

me. [...]

It was a great relief for me to hear specifics, in contrast to (comic) generalities about (dis)likes. They definitely fitted my own views of the paper.[6] (As I said I have liked such things too.)

28 July 2003 to Robert Olby. Brecht's *alienation* is familiar (in plain English) from such commonplaces as: Don't make it too easy for the reader, tempting him to skip. Who am I to tempt you? Half the world (I suppose) prays every day not to be led into temptation. [...]

Crick says, quite correctly, about me that I am ponderous except when malicious.

24 August 2003 to Kenneth Derus. In my early teens the first theorem in geometry that aroused my interest at all concerned the angles subtended by a segment of a circle at its centre and at points of its circumference (on the 2 sides of the segment). About a couple of years later—this is the anecdote—quite spontaneously I had the feeling I had NOT BEGUN TO UNDERSTAND the theorem (although I knew statement and proof), because it had not occurred to me to ask (myself or anybody) WHAT was interesting about it. At the time I persuaded myself that one could NOT VISUALIZE the theorem, and NEEDED A PROOF (which earlier theorems had not needed). NB I certainly knew that the angles subtended at points on the circumference were different (on the 2 sides of the segment except when it was a diameter), but had not related it to (not) visualizing the theorem near the end points of the segment.

14 October 2003 to Kenneth Derus. *Reminder* around Goethe's remark to Eckermann about a guiding (general, aka abstract) idea in *Faust*, tacitly, in 6.V.1827; after it was completed. Goethe emphasized the *richness* of the material. 20 years earlier (in 1806) Goethe rejected the impression of Heinrich Luden, quoted in Biedermann, that Faust, as conceived, lacked any such idea; roughly, as Goethe put it modestly it was not just a string of pearls. I'd say there was the *literary idea* of *chitchat with the devil*, where there are no holds barred.

---

ing INTELLECTUALLY NEW, but CONFIDENCE IN THE OTHER'S UNDERSTANDING, which is occasionally rewarding. *GK*

[6]See [Kreisel, 1992].

17 January 2004 to Kenneth Derus. I vaguely remember your interest in some lectures (at Stanford, 1969) on *Algebraic Proof Theory*, which (as I see them) came still born off the copier. – The title fits better a paper by Baaz & Wojtylak (at least as I have come to read it), [BW] for short,[7] where formal rules are classified according to the kinds of WORD EQUATIONS associated with the proofs generated by the rules; for free semi groups with one generator in the case of monadic systems (= only one function symbol required to be monadic), unification and semi unification (with non-trivial (semi) unifier when the rules are *not* monadic). – The authors have asked me to write a PS to [BW] (and I hope to start on it soon).[8]

17 February 2004 to Kenneth Derus. Perhaps the typed PS will sound different from the MS for me, too. It will be refreshing to have comments in your usual incisive style with memorable points, especially in contrast to the pieties of Baaz, let alone Wojtylak.

18 March 2004 to Kenneth Derus. *Reminder* After mid (or, realistically, the end of) April I'd also be interested in your comments on the combination of PS [BW][9] and *Around the metaphor* ...[10] Needless to say I do not assume that you share those particular interests. It's just—as I mentioned before—most of my more playful interests before the episode of clinical depression have not survived the latter.

In connection with the projected article *Bourbaki's Foundations: (re)viewed after 50 years*[11] other points will turn up that may have an interest we share.

14 April 2004 to Kenneth Derus. I have been reading LEIDEN UND GRÖSSE DER MEISTER by Thomas Mann (Fischer Bücherei, Frankfort 1957). Naturally, it is a bit ponderous. (Einstein felt, when Mann was a neighbor at Princeton, that he might get an explanation of relativity any day.) [...]

---

[7]See [Baaz and Wojtylak, 2008].
[8]PS to [BW] = Kreisel's 5,990-word text. Cf. [Baaz and Wojtylak, 2008, pp.130–38].
[9]PS to 'Generalizing proofs in monadic languages' (16,971 words, unpublished). (PS [BW] ≠ PS to [BW].)
[10]Around the metaphor of logic as—the grammar of a—language for abstract mathematics (9,463 words, unpublished).
[11]Bourbaki's foundations (re)viewed after some 5 decades (16,437 words, unpublished).

I once asked McGuinness, a biographer of Wittgenstein, to tell me a single aperçu of Wittgenstein, which is (a) *not* specifically related to logical foundations, and (b) *not* found in Goethe already.

29 June 2004 to Kenneth Derus. *Reminder* of Tait's counterpart to the article in *Synthese* 114[12] (from which you sent me excerpts). He genuinely contributed to that paper (in 1961); specifically, though the logic (which I had done) established the general form of the solution set, Tait worked out an explicit form (by algebra). Admittedly, I wrote the published version (and did not try to test his understanding). I am sure he was never receptive of the kind of subordinate clauses to which I pay attention. *Remark* Tait's favourite literature was not Thomas Mann, but Tolkien's *Lord of the Rings*. (Harry Potter was not available then.)

2 July 2004 to Angus Macintyre. In the article I wrote at Oxford in memory of Myhill I described my conversations—actually, not only—with him as soliloquies in the presence of another. Most of my letters are soliloquies with another in mind, and so do not require replies; any reply they evoke is thus a bonus.

16 July 2004 to Hubert Faure. About your question, whether I have ever felt afraid of the future. If taken literally (meaning a conscious feeling) the answer was easy (and correct): No. But there is a variant, which is (for me) of very real consequence: Do I have reasons for being afraid of the future? You know as well as I that I do. [...]

PS around ideals of social chitchat (which is, after all, the principal arena of your intellectual experience). Scintillating conversation—in French, English, American and other styles—has room for shallow and superficial remarks (here meant in senses understood outside social traditions).

23 July 2004 to Kenneth Derus. Around 1980 I spent the weekends at Lacan's house (near Guiterencourt) while visiting the IHES at Bures. His—usually totally silent—presence had a very calming effect on me. (He said that, in his experience psychoanalysis served well for literary criticism, not for medical purposes. He had many patients, who were prepared to pay high fees.) Whether or not one is attracted by his theatrical manner, I had the impression of high intelligence.

---

[12] See [Kreisel, 1998].

19 July 2004 to Daniel Isaacson. Plenty of people were impressed by Wittgenstein's long pauses in his seminars. I have never had any objection to long pauses, but if one considers what he said at the end one really wonders whether *that* required such pauses.

6 August 2004 to Kenneth Derus. I have remembered a—what I have always felt to have been a—memorable experience > 30 years ago.

It was a strange psychiatric experience of NOT—having the feeling I'd describe as—HAVING A WHOLE PROOF BEFORE MY EYES. (I probably mentioned it to you.) I warned the students before I lectured (nevertheless). As I went along the steps presented themselves in time. *I* found this activity disagreeable. The students found the lecture 'clear'. – It turned out I suffered from a low grade glandular infection.

7 August 2004 to Kenneth Derus. As to *sentimentality*, I have been accustomed since my teens to pay attention to differences between feelings and their (verbal or other) expressions; I was struck by the clumsiness of most, including my, expressions of feelings in contrast to those of good actors with a script by talented poets. In particular, there was the matter of sentimental expressions (with a kind of corollary about hackneyed, mechanical feelings fitting those expressions). *Anecdote* from my teens (vaguely related to these matters). Once, after I had made one of those remarks to Wittgenstein that have come up in our correspondence he told me that, if ever I committed a murder, he was pretty sure that (a) he'd have no objection to the deed, but (b) he'd hate the way I'd talk about it.

3 September 2004 to Kenneth Derus. PS about Feyerabend('s autobiography). Actually, Crick drew my attention to a reference according to which the eldest daughter of Geach and Anscombe reported experiences (memory of abuse and abuse of memory). Crick took it at face value. Certainly, there is some element of truth in it. The children of the philosopher pair, especially the only son (John Richard, whom I have quoted in print) at the age of 5, decided that, if they felt the parents did not understand this or that, they'd come to me for answers. (At the time Miss Anscombe was translating Wittgenstein, and I visited occasionally.) The daughter had overheard the mother *talk* about particular kisses, but then refused to give details when asked for them. It would have been quite unrealistic to teach the daughter by demonstration.

21 September 2004 to Kenneth Derus. Thank you for your emphatic e-mail of 16.IX.04 on—the (in France) notorious flatness of —Dieudonné's style. [...]

Anecdote about Thomas Mann. Everybody knows that the (original) translations into English are pathetic. Somebody commiserated with Thomas Mann, who replied to the effect: She is such a nice woman. Why should she know German, too?

2 November 2004 to Kai Käkelä. In reply to your letter of 14.X., let me say that I hold no brief for your project of reviewing— material like—Gödel IV and V, given your limited experience of men and events. But your letter may reflect impressions and views of others, too, and so is of interest beyond only one person.
1. A couple of corrections. (a) If not Prof. Cohen I did give the editors—what I still regard as—a reason for not publishing my correspondence with Gödel in their collection; cf. (3) below. There was no reason for them to publish it. (b) Despite your professional credentials you do not seem to master the (familiar) distinction between personal (likes) and non-personal (reasons). So I shall not rely on it below. NB (2) is more detailed than the reason given to the editors.
2. My main contact with Gödel was in private conversations during the years I spent at the same Institute, not in correspondence. What is more my letters were not intended for publication, and so referred often, and usually tacitly, to points that had come up in conversation (or were to be taken up in later conversations). So an edition would have to make those tacit understandings explicit. Nevertheless, the correspondence was extensive, and so a selection was required.

Besides, the correspondence was in German, and would have to be translated. I was not prepared to do all this necessary work myself; cf. (3).
3. I do not have the intellectual confidence in the editorial team to rely on their help. This is an open secret; cf. the many differences in interpretation in the published literature; most recently in *Synthese* 114 (1998) 99–160, and references throughout. Note 1 on pp.149–150 gives a list of reports on Gödel since his death.

7 December 2004 to Kenneth Derus. As to the 2 trifles in Eureka 1944,[13] I remember they were written in 1943. (I was struck by the quick publication

---

[13] See [Kreisel, 1944a] and [Kreisel, 1944b].

in the middle of the war.) I went up to Cambridge, as one said then, in Jan 1942, an irregularity, also related to the war.

Even though I did not know the (Jesuit) principle (going back to Ignatius of Loyola himself) I practiced it by *not* stating the logical idea(l) followed by Littlewood in his proof of Schröder/Bernstein (of avoiding reference to the natural numbers), while, of course, Hilbert/Bernays vol. I does just the opposite. Oddly (in the other trifle) I used the then-current expression: THE necessary and sufficient condition, tacitly, where (mere) logical equivalence is understood in the definite article; oddly, because in some examination in 1943 there was a question asking for THE necessary condition for $\partial^2 f/\partial x \partial y = \partial^2 f/\partial y \partial x$. I had used up the time for the rest of the question, and gave as—in words to this effect—a CORRECT, BUT NOT DESIRED answer: $\partial^2 f/\partial x \partial y = \partial^2 f/\partial y \partial x$. I have no idea if marks (as one said in Cambridge UK, perhaps, grades, in Cambridge, Mass) were given for it; I am sure, at that time there was no penalty for (such) frivolity.

13 December 2004 to Angus Macintyre. (a) Feferman likes to be *judicious*, but simply does not see elements that stare others—as always, with suitable BOR[14]—in the face, besides contradicting himself—if not in the next sentence, then—in the next paragraph. Let me take an example from *Kreiseliana* on the subject of my (not) being precocious. (i) I was supposed to be slow (if not as slow as he) by going up to Cambridge, where he assumed it was October 1942. In fact, it was Jan 1942 straight after the scholarship exams for Trinity in December (then). (ii) I was supposed to be mature because I had read Hilbert/Bernays before I was 20. (b) Admittedly, Lighthill's obituary appeared after *Kreiseliana*. But even if Feferman had read it, he wouldn't have remembered that Dyson and Lighthill worked through vol. 1 of *Principia Mathematica* at age 14.

As to Tarski, when I met Feferman in summer 1956 (at Berkeley) he felt ill treated by Tarski, who refused to accept the dreary business later published under the title *Metamathematics in a general setting* (and sent him to Henkin for dissertation advice about an even drearier business replacing familiar results about recursively inseparable r.e. sets by showing them to be creative). I told Feferman that *if* one wants to fuss about formulations of consistency statements one had better look at Turing's ordinal logic, where such statements are iterated transfinitely and so attention to detail is rewarding.

---

[14]BOR = background and other resources.

15 December 2004 to Mark van Atten. Hintikka has a record of getting hold of the wrong end of whatever stick he happens to grasp (and to thrash it around). For example, he attributed autism to—of all people—Wittgenstein, one of the touchiest souls I have come across.

I once reviewed a paper of his, in which he described himself as *seelenverwandt* with Gödel.[15] If I am not mistaken the review was Zbl. 820.3001.[16]

17 December 2004 to Kenneth Derus. When Isaiah Berlin visited Husserl they went together to a shop where his host regularly bought tobacco for his pipes. Before they entered Husserl whispered to Berlin: '*Sie wissen nicht, wer ich bin*'.

3 January 2005 to Daniel Isaacson. I see no reason to encourage amateurs simply because they love what they are doing (Macintyre says that I don't disguise my feelings); cf. Trollope (Ayala's Angel): It has been my observation that when people are interrupted in some egregious stupidity, they always feel hurt.

5 January 2005 to Angus Macintyre. It is always useful to cite Gödel for a catchy opinion; if nothing else, for its preposterous exaggerations: the cardinality of $\mathbb{R}$ is the most fundamental property of the reals. As we know (by TENOS[17]), it's hardly ever of consequence in mathematical experience; tacitly, apart from being $> \aleph_0$. [...]

Of course, there are most gifted, but inarticulate mathematicians (at least with casual conversation partners), for example, Langlands. (I may have told you about my chats with him at Bures.) But there was nothing vacuous even in such casual conversations.

19 January 2005 to Kenneth Derus. In the 1960s when I engaged in social life I liked the company of actors who didn't stop acting also off duty. Come to think of it, I was very fond of Wittgenstein's company who complained in his so called secret diaries of feeling that he was constantly acting. *Reminder* I giggled throughout his seminars; not because he was acting sitting in the posture of Rodin's *Penseur*, but because it was clumsy. During walks or generally *à deux* it struck me as impeccable.

---

[15] Cf. (for contrast) *Synthese* 114, p.150, l.9–16 where I document ways in which I was not a soul mate of Gödel's. *GK*

[16] See [Kreisel, 1993].

[17] TENOS = tested experience not only speculation.

25 January 2005 to Angus Macintyre. As a father you know better than I if it is true that babies present their daily excretions as products of obvious interest to all around. But as a metaphor it is good enough to be prepared for the innocence with which, e.g., Frege presented his answer to: What is the number 1? Admittedly, he may have preferred Brouwer's attacks to my defense of his doings.

14 February 2005 to Kenneth Derus. For nearly 6 decades I have been aware of (and have stressed) the possibility of regarding the kind of proof theory I did as refining suitable model-theoretic results; in contrast to such things as the consistency business in cases where the consistency (of the principles considered) is model-theoretically a matter of course. *Reminder* To the proof theoretically ignorant—and thus if sensible uninterested—this practice is liable to consolidate lack of interest: what they know of the refinements will certainly be of unsatisfactory marginal utility compared to—their knowledge of—the model-theoretic results in view. NB The rule of thumb above, popularly aka philosophy, is a nice idea, to be remembered, not dwelt on. One will find *other* refinements. [...]

They say: *Tout comprendre c'est tout pardonner*. True or not, I don't feel it concerns me (as it didn't Catherine, the Great, on her death bed when asked to pray that God may forgive her: *C'est son metier*).

22 February 2005 to Kenneth Derus. *Remark* on the popularity of worldly wisdom ranging from the Jesuits to Gödel who shared the dislike of them which was in Austria conventional. Another item of worldly wisdom dear to him—and in conflict with the Jesuit principle of never refuting, only embarrassing opponents—was to attribute sensible views to those felt to be opponents: Either they really held that view without having expressed it properly or, if not, one is at least not left empty-handed.

1 March 2005 to Kenneth Derus. *Anecdote* In a footnote (Bull. A.M.S. 84 (1978) 79–90) I quoted a quip by Dyson (to Wittgenstein when Dyson was 18).[18] I thought the quip was sensible, but Dyson thought of it as a youthful indiscretion, and wrote back that Wittgenstein was like Oppenheimer: when they said something completely banal, people quoted them with awe. I replied

---

[18]Cf. [Kreisel, 1978, p.86]: "Dyson had said he did not wish to 'discuss' anything, because *what* W had to say was not different from anything everybody was saying anyway, but he wanted to hear *how* W put it."

that I agreed completely, and reminded him that I was quoting him, not Wittgenstein. (Dyson was annoyed.) *Remark* Also at the age of 18 Dyson told me Cambridge gossip about a (poor) mathematician Forsythe, who occurs in *Littlewood's Miscellany*. Forsythe was a bachelor, and the wife of another mathematics professor appeared late at night with a suitcase at his house. Apparently, Forsythe resigned his professorship, married the lady, and was unhappy ever after.

7 March 2005 to Kenneth Derus. Very many thanks! Wittgenstein's twist on Frege's 'Einbruch der Psychologie in die Logik' has the charm I remember from our conversations. *Reassurance* In the case of Frege the last thing to look for is humour. He tries to make 'telling points', which then say more about him than—the usually obvious—points in view.

22 March 2005 to Kenneth Derus. Before I went up to Cambridge early in 1942, I slept through air raids; once a piece of antiaircraft shrapnel fell through the roof of my bedroom without waking me. After 2 years at Cambridge where it was exceptionally quiet I was a light sleeper, easily disturbed by sound and light.

5 May 2005 to Matt Ridley. Crick and I attended a lecture by Dirac in the 1940s on magnetic monopoles, with a (to me) memorable use of topological aspects of Dirac's equation. Crick was most struck by Dirac's manner: of a grammar, not a public school boy. Crick didn't know the—then, perhaps no longer—familiar joke about *differences* between Old Etonians, Wykehamists and Harrovians (at a cricket match and a lady in need of a chair).

14 May 2005 to Angus Macintyre. After Gödel's proof was explained to Wittgenstein he made it palatable to himself by considering (after Gödel, but before Henkin) a list of rules $p_1, p_2, \ldots$ and 2 rules p_G and p_H on that list, saying: write $p_G(n) = 0, 1$ if $p_n(n)$ tells you to write $p_G(n) = 0, 1$ if $p_n(n)$ tells you to write $1, \neq 1$ and $p_H(n)$ the same as $p_n(n)$. Quite rightly (tacitly, for formal rules in the literature), he concluded that you must not write anything at $p_G(g)$ if p_G is the gth rule on the list, but also that you may write anything (or nothing at all) at $p_H(h)$. This second conclusion is valid only with fake good will: there are formal rules, for which all Henkin sentences are provable, there are others (for example, certain cut free ones) for which some Henkin sentences are provable, but literal ones refutable, and yet others where—of course, some are provable, but—literal ones are undecided.

Disclaimer I do *not recommend generally* paying attention either to—the likes of—the Liar business or to attempts by the uneducated to make professional knowledge understandable without professional resources.

6 June 2005 to Matt Ridley. You did not tell me when Crick joined Jensen and Shockley on the matter of inheritance of intelligence. As it happens, for quite a time I lunched regularly with Shockley at the Faculty Club. He was at Stanford, having left Bell in the 1950s to make money for himself with several ideas he had for developing semiconductors. At Stanford people avoided his company, at least in public. For me he had a particular charm. Other people, who keep diaries, might be satisfied to note down such stories as below privately; I have never kept a diary, and so it is more natural to put them in a letter.

13 June 2005 to Matt Ridley. At another extreme there was Watson's *Double Helix*. Before its publication Crick asked me—I still have not given a thought to whether he was serious—whether I would join him in a legal action. Again I merely quipped to the effect that, whatever he may have said about me I am sure he could have said worse. *Remark* In retrospect, after reading Watson's book I realize that a court case would have been an occasion for fun (tacitly, for those with a gift for enjoying public cases). As least as I read it Watson gave the impression of wanting to be frank, but omitted all those antics of his that made it difficult for many to treat him as an adult.

13 June 2005 to Newton da Costa. I have heard about the would-be biography of Tarski by the Fefermans. I have not looked at it. *Anecdote* Soon after the revolution in 1917 there was a conference of the *Internationale*, to which a Polynesian delegation came. The USSR was disorganized, and no translator could be found. According to my supervisor (Besicovitch, at Cambridge), who was then in the USSR, Radek, later murdered by Stalin, offered to translate simultaneously. He spoke fluently. Afterwards Lenin mentioned something to the effect that he had always realized Radek's broad education, but not his knowledge of Polynesian dialects. Radek demurred, and Lenin asked how he knew what those Polynesians said. Radek's reply (in Russian): What could they have said? – If you think about it, you'll see how it relates to Feferman's stories about me or at least, why I don't expect to learn much there.

24 June 2005 to Angus Macintyre. Lang. According to Serre, Lang, Lacombe, he and I were together somewhere, sometime in the 1950s, and I said something that disturbed Lang enough to leave for the lavatory to be sick. (Serre assumed I'd remember the occasion so there was no need for further details.) Now, I remember quite a lot about the 1950s or even 1940s; cf. my letters to Ridley. But though the stories that Fernau and Gandy remember about what I am supposed to have said ring true (in my ears), most of them concern aspects, generally, with the socalled human touch, that I don't remember.

12 July 2005 to Grigori Mints. *Anecdote* Herr Weingartner finds that everything I say is surely known by those to whom I say it, but what they say sounds as if they did not know it.

26 July 2005 to Kenneth Derus. Some biographers have, I feel, made excellent use of SELECTED passages from letters (admittedly, by excellent authors); for example, Pais in biographies of Einstein and Niels Bohr. NB I realize that I can't expect to have similar feelings about mine (quite independently of considerations of modesty). My overwhelming feeling about my own letters is this: I have seen it all before.

22 September 2005 to Angus Macintyre. At least for my understanding of the many brash announcements of Gödel and Weil there is a striking difference. Gödel's are really brash, if not barmy (but in his good days tempered by nice, sharp points; cf. my review in NDJFL of vol.2 of his *Collected Works*). Weil's are (as I emphasized above, to me) possibly affected, but sensible, even commonplace. *Reminder* When asked for his principal rule of thumb he chose: *élargir votre catégoire*. Well, take a broader view is pretty venerable; tacitly, of course, not just any old, but a suitable broadening. *Remark* As they say, 'one knows what he means', but then one knew it before he said it.

23 September 2005 to Kenneth Derus. The story as told is pure fantasy. I never took Mrs Dyson, with or without those 2 children, to tea with Gödel. Nor did she see Gödel (or his wife) in Spring 1957, when I myself was in Princeton. She did once go to tea with the Gödels, possibly taking the children along, while I was *not* in Princeton; I don't remember if I was back in England or elsewhere. That meeting was my 'doing' inasmuch as I wrote to Gödel suggesting he should see her in order to see with his own eyes a then-popular subject of gossip at the IAS. (He and Oppenheimer decided to invite me as a

Research Associate and not a Member because research associates (of a professor) were invited by just the director and the professor concerned, while members were invited by the Faculty.) It won't surprise you that—in contrast to Bertrand Russell, who was indignant about the City College of NY—I was quite satisfied to see formal difficulties solved in formal ways.

28 November 2005 to Mark van Atten. As to your meeting at Lille, I recognize of course the friendliness of your invitation. *Remark* about others of my age. Many of them have retained their physical stamina, like to travel (and to be honoured guests), and do not get tired of repeating what they have been saying over the last decades (if not longer); possibly with the feeling of thereby enlightening the young. I do not have any of these gifts.

6 December 2005 to Juliette Kennedy. *Remark* on a certain sobriety on Gödel's part at the time. In Brno he had a girl friend who conveniently lived in the same house. When she expressed impatience with his valetudinarian pre-occupations by saying that he must choose been her and his stomach he (reasonably) pointed out that he had no choice.

6 January 2006 to Kenneth Derus. What, if anything, could be done better (and hence differently) in biographical memoirs for Brouwer, Russell, Gödel after reading Janet Malcolm's book. *Reminder*. It's difficult for me to remember the energy and concentration I had 40 or even 25 years ago, when I wrote those things. Certainly, I know a fair amount of (relevant) logic now that I didn't know then (and should be glad to include now, at least in asides). Also I couldn't have resisted the temptation to make a similar joke about the collaboration with Whitehead (by lubrication?), if I had known that Russell had an affair with Whitehead's wife; cf. the—35 years ago daring—joke about Delacroix (at the Quai d'Orsay) and Talleyrand. But how could the 3rd abstract part of *The silent woman* have helped?

19 January 2006 to Angus Macintyre. Wittgenstein bragged about being a professor of philosophy without ever having looked at Aristotle. (If he had he'd have found many things there that he said—admittedly in his own way—himself, confirming his complaint (about himself) of never having had an original thought in his life; he even knew the reason (of non-Aryan descent, but becoming an honorary Aryan by Göring's personal intervention, when a large part of the family fortune was transferred to the Reichsbank; only the part belonging to his brother, the one-armed pianist, who lived in NY was not transferred).) [...]

Quite recently my impression of Aristotle wanting to 'do down' Plato— or, putting it more positively, confining himself to points that Plato had not elaborated before—was confirmed by Plato's own nickname for Aristotle: the *foal*. A tacit understanding at the time was that a foal kicks its (mother) mare when the milk is gone.[19] (I don't know if this is generally true of foals or, for that matter, of whelps or kittens, mutatus mutandis; they'd bite or perhaps piss on their mothers rather than kick them.)

27 January 2006 to Kenneth Derus. Presumably you have read the autobiographical part of Broad's contribution to Schilpp's volume on his philosophy. His self-confessed shyness fitted his conduct when he was Junior Bursar (at Trinity, Cambridge), and had the possibility of abusing his position, when attractive undergraduates misbehaved, but didn't. – I don't think I can convey the full inwardness of such situations.

7 February 2006 to Kenneth Derus. I often don't know very well whether I am writing to you or to Macintyre.

7 February 2006 to Angus Macintyre. *Reminder* of the popular idea(l) of *adding the human touch*; above, of Gödel's enthusiasm for the lectures on class field theory (where the Chinese remainder theorem was used, but presumably not in relation to cohomology) by Furtwängler, who, like Hawking, lectured from a wheel chair, but because of being incapacitated by the pox (of 2nd degree) instead of a molecular disease. On balance I feel the human touch wastes time if one has something interesting to say (of course, almost as a corollary, it can be a Godsend if one doesn't).

13 February 2006 to Angus Macintyre. PPS about Odifreddi. At one time he proposed to take me (of all people) under his wings, and made some agreement with del Franco (of Bibliopolis) of *editing* of volumes of selected essays etc. by me. (I was actually quite curious to see what he'd keep in and what he'd edit out or at least 'curtail'. I told del Franco *my* interest, and noted that this

---

[19] At the age of 6 (presumably without knowing all this) Esther Dyson reacted to complaints about her mother abandoning the family with the question: What use is a mother when the milk is gone? *GK*

was not necessarily in the public interest. I don't know if he took it in.) Without much exaggeration Odifreddi seemed taken with my formulations of quite familiar points simply because they were not as heavy as Feferman's or Troelstra's. I have no reason to assume that he afterwards collected *Kreiseliana* as revenge (for my not agreeing to the publication of his 700 pages or so of editing). Not all people who contributed to *Kreiseliana* did so to get 'even' with me (or in the case of Mrs. Dyson with him); some just were angry. (Some were suspicious, and refused altogether; e.g. Iris Murdoch and Hubert Faure (with the feeling—they told me—that no good would come of it).)

16 February 2006 to Kenneth Derus. I knew Zaehner at All Souls in the late 1950s. Fairly regularly he got drunk in Warden Sparrow's rooms. Once Zaehner's (thick) spectacles slipped off his nose and he trod on them. Next morning Sparrow sent them to Zaehner's rooms with a note: 'This morning I was greeted by this sorry spectacle'.

8 March 2006 to Kenneth Derus. I had not heard of Molyneux's question, which is more congenial to me than a good many popular questions of contemporary would-be experimental psychology I have come across. [...]

Generally, about—your interests around—consciousness. I suppose it would be fair to speak of MAMNE's view,[20] but informed by considerable BOR.

12 May 2006 to Angus Macintyre. Admittedly, in the 1940s I was dubious about—the axioms GB for—sets, and called the result that GB had no recursive model *hypothetical* (as an example of a consistent formulae without a recursive model). In the early 1950s I came across the full cumulative hierarchy of sets, and have seen no reason to doubt its *definiteness* provided it is built up from a *single* atom, and is bounded by a suitable uncountable strongly inaccessible cardinal (with the classes being first order formulae).

24 May 2006 to Dana Scott. *Remark* on my—not (on my side) un-friendly, but simply—nonexistent relations with the enterprise of publishing Gödel's *Collected Works*. Many years ago Feferman told me he had, tacitly, heard clearly, but not understood some explanation I had given. When I gave it again he told me he did not understand what he had not understood the first time. It

---

[20]MAMNE = many a mind's naked eye.

seemed sensible not to distract him from those with whom he communicated more easily.

26 May 2006 to Dana Scott. When Cohen's twins asked me (to my surprise) if I had a message for their father I said I should not dream of formulating one spontaneously if he was still exceptionally touchy, as he was in the 1960s when I had contact with him, and Eric said he still was. – *Afterthought.* The twins asked me about my first meeting with Cohen, which actually left no lasting impression on me. The first meeting I remember was a dinner at some restaurant in El Camino, where I asked him about Weil's famous use of Lefschetz's theorem for proving—a suitable topological interpretation of—the Riemann hypothesis for function fields. Cohen, who of course knew of Weil's work (and according to Gödel had impressed Weil as the best student Weil ever had), said he didn't 'understand' that use; tacitly meaning—what Cohen would have felt to be—a mathematical core in terms of a classical tradition.

Being self-educated is not a total loss, though it usually involves gaps in common knowledge among *bona fide* professionals.

26 May 2006 to Matt Ridley. *Anecdote* When I was in Paris I saw regularly Queneau, a successful novelist (*Zazie dans le métro*) who was fascinated by mathematics (and I suppose some mathematicians), but rather more critical nearer to home. He put it this way: There are good, bad and indifferent mathematicians, and many are not mathematicians at all. But who is *not* a novelist? (I came to know him through the novelist / philosopher Iris Murdoch, who began to write novels at the age of 12, and kept them in a drawer. She told me she published some of those early novels after she had become successful.)

1 June 2006 to Angus Macintyre. Certainly, Wittgenstein, but also Gödel, and perhaps also André Weil had somewhat lopsided gifts; in contrast to, say, Serre or Deligne. (René Thom's gifts struck me as lopsided, too.) – In real life this did not at all reduce my liking for their company, which, for all I know, may have been a welcome change from the attitude of others. ('They' are here the 'severely' lopsided; André Weil's nervous manner did not suit me at all.)

6 June 2006 to Brian McGuinness. Wittgenstein occasionally forgot that he could be boring. *Reminder* (of an anecdote in print). I pointed out that he was repeating himself, and he tried to console me that it was still true if it was true before. I agreed of course, but added that even if it was not boring the first time it certainly was boring now.

24 August 2006 to Angus Macintyre. Overleaf you find a copy of comments by Derus about the drafts of my Templeton article he had typed (so far); more precisely #1–58 are about Sections 1–3 and corresponding endnotes, #59–67 about discarded variants for Section 4. [...]

It is surprising to me that somebody who is really not a professional logician at all would pick out items that I myself feel worth remembering. Whatever his reasons may be, in effect his list is of interest to me; very much in contrast to the people who tell me (so non-commitally as to sound evasive) that they 'like' my letters or are 'interested' in them. *Remark* It wouldn't surprise me if those—not only evasive-sounding, but—decidedly slow witted people (whether they twitter rapidly, or ejaculate short would-be judicious pronouncements) would overlook that I have said very little about *my* interest; perhaps, not even—what I said about—my surprise. But, of course, just because Derus has given me so much help, his comments provide, in effect, also reassurance (about his feeling *he* has got something out of all this, too).

14 November 2006 to Brian McGuinness. I never asked Wittgenstein (or others with rare exceptions) to elaborate; after all, if people are clever enough to elaborate sensibly (and know me) they could be expected to know that I welcome such elaborations.

22 November 2006 to Brian McGuinness. Something that—perhaps not for your idea(l) of your article, but—for those not yet exhausted by tales around Nazi antics could be of interest and is not pursued in your article is the style of the Wittgenstein family's dealings with legal problems (in addition to what struck Hermine about her family's self-assured manner before judges etc.): When one member of the family—I expect you know his name—was caught with a fake passport and a forged signature on the Yugoslav frontier, the family employed the lawyer's firm of Seiß-Inquart, who later became Reichskommissar of the Netherlands, and was executed at Nuremberg. Somehow, the accusation was (only) for forgery, and the defense argued that this did not apply to a signature on a fake 'document'; successfully. For all I know, at the time, any judge would think twice about rejecting a defense by Seiß-Inquart, and any prosecutor before appealing against a judgment that accepts such a defense. [...]

Miss Anscombe (told me she) asked Wittgenstein if her husband would be welcome in Wittgenstein's seminar. She was impressed by his condition that Geach should not bring up examples from Babylonian mathematics.

5 December 2006 to Kenneth Derus. I just got your (very) welcome e-mail dated—as usual, here—4.XII.06. – Just compare the would-be precise formulations by Watson[21] with civilized comments on Gödel's theorem, for example—since you have LH at hand—in our piece for the Templeton volume.[22]

7 December 2006 to Daniel Isaacson. When I went off my head 3 years ago, the only memorable *personal* experience I ever had, I was terrified of just this kind of loss of proportion and of ponderousness. When Derus sent me typescripts of what I wrote during that period, there was certainly a striking reduction in quantity, but no dramatic difference in quality.

12 December 2006 to Michael Dummett. *Anecdote* about Kripke (which I may have mentioned though I don't remember doing this). When he was 12, at a summer camp for children of this age, he was thrilled by Hume, especially the recommendation, which certainly applies to most of Hume's writing, too, to the effect that books should be thrown away unless they contain empirical facts or mathematics (or, at least, something like this). He was so excited that in the middle of the night (when he happened to read this) he went round the camp, waking up the boys to read Hume to them. According to Kripke the boys were not grateful. *A parallel* (?) When Wittgenstein was a (modest) village schoolmaster, he went with some visitor from Vienna to the village pub, full of locals. In his high-pitched voice he said: Ich hatte einer Diener namens Konstatin (and, as he told me, he was taken aback when the whole pub stared at him; in his usual way he went on at length about diverse interpretations of the pub(lic)'s reaction).

19 January 2007 to Angus Macintyre. PPS about Cohen's pronouncements on Skolem's contribution. I hold no brief for Cohen's formulations. But

---

[21] See [Watson, 1938].

[22] LH = Logical hygiene, foundations, and abstractions: diversity among aspects and options. 26 drafts of this text exist. The published draft excludes Abstractions related to mental processes (the 11,904-word last section). Cf. [Kreisel, 2011].

Gödel's own claims, cited by Kohlenbach, of the superiority of his own philosophical views over those of Skolem and Herbrand are farther off the mark than Cohen's. As to formulations, admittedly, I have been—spoilt by—reading Tao, as articulate an exceptionally gifted mathematician as Serre or Deligne.

5 February 2007 to Angus Macintyre. PS about my exercises around $\pi(x) - li(x)$ in the latish 1940s (published in the early 1950s). There is a difference from Kohlenbach's expositions: his can be understood if one knows *only* functional analysis; he just mentions that this bit was inspired by a logical metatheorem or that is an instance. I on the other hand made a point of interrupting the analysis (after all proofs of analytic number theory are mostly analysis, only the theorems come from number theory), and emphasized local relations to logic. So, a reader had better know *both* a fair amount of analytic number theory *and* at least a little proof theory.

14 February 2007 to Kenneth Derus. Actually, not only you seemed to like the article in *Gödel Remembered*; by a fluke, 2 people whose praise would have made me very ill at ease (Dawson and Feferman), —in effect, though presumably not on purpose—went out of their way to reassure me by expressing their discontent.[23]

15 February 2007 to Kenneth Derus. At the risk of repeating myself, let me say a word about some differences *between* your selections of snippets from our correspondence *and*—what I regard as—reactions to my letters by others; including Macintyre, Kohlenbach and, at the other end of the scale, Baaz or Faure. It's not only that *I* get nothing from their reactions, but I at least do not find any indication of anything *they* get out of my letters (though they are all polite about them). Oddly, in the case of Mints, there is occasionally a trace of his using my letters. *Disclaimer* (where I am sure I am repeating myself). This lack of reaction by those addressed in my letters is no reason for my *not writing* them; I expect that some of the snippets you sent me earlier this week are from letters to third persons.

21 February 2007 to Kenneth Derus. Your # 8 was a revelation, specifically, of a side of Stanford in the days when it was said to be a Playboys' university. (During my first visit at the end of the 1950s it still was, and I, for one, liked this.) – In those days I saw a certain amount of (Richard) Montague,

---

[23] See [Kreisel, 1987].

and his (pretty) rich pieces of trade, of whom he was as proud as Waldburg/Zeil of his wife (at Hohenems; he is a—rather typical—Habsburg, she comes from the Bavarian branch of the Schönborns, her father, a general, was executed after the failed generals' putsch on 20.VII.1944). Probably it says more about me than about them that I have remained so impressed by their pride. In any case, I found the whole story delightful. Whatever your measure, by which quotations from my letters make up 40% of your letter (of 13.II.07), in my memory the material in small print makes up > 90%.[24]

Your # 12. Several French readers, including Faure, expressed enthusiasm (to me) about Poe that was unreal for me, without it ever occurring to me that the translation was an improvement. (It 'figures'.)

23 March 2007 to Kenneth Derus. Macintyre just called; not quite as exuberantly (or perhaps better effervescently) as the announcement you sent me about E_8, but in this direction. – At any rate, when brought back to something manageable (Dyson and partitions) he was able to complement—what you said in your pdf dated 18.III.07 about—relations between weak Maass forms and l-adic representations.

7 May 2007 to John Chancellor. *Reminder* of Ridley. He certainly did not hesitate to include stories (others not I had told him) about me in his biography of Crick, for the human side as it were, but again it wouldn't surprise me if he, too, disapproved. Before publication he showed me the book, and asked me if I liked what he had written about me. I quoted to him Esther Dyson at the age of 5: I don't have to like it to be interested.

23 May 2007 to Kenneth Derus. As far as—enjoying—Feyerabend's *Killing Time* is concerned you have a great advantage over me by *not* having stayed with the Geaches: compared to the real thing Feyerabend's descriptions (albeit no doubt well written) are a very pale second best.

25 May 2007 to Angus Macintyre. Not everything spectacular is mathematical(ly interesting), not even in mathematics; cf. Feynman's use of Newton's general laws.

---

[24]Material by and around Kenneth Rexroth and Yvor Winters. Cf. [Rexroth, 1981, pp.19f] and [Winters, 2000, pp.323f].

20 June 2007 to Angus Macintyre. (a) On the one hand, —it seemed to me, and perhaps also seems to you—having emphasized the formalization in PA of Wiles' proof more had better be said about it than even in your second draft. (b) On the other hand, at least to me (possibly not to you) it would go against the grain not to mention any, at least temporarily satisfactory corollary to this. Well, by cut elimination, at least since the 1960s it has been known that proofs of Π_1^0 theorems in PA can be transformed into proofs by Fermat descent (quantifier-free transfinite induction with quite elementary descent functions).

21 June 2007 to Kenneth Derus. For all I know Friedman is one of these few intelligent—among very many, let us say, not so intelligent—people who are deeply convinced they have broad and sound views without ever having put this impression to the test. There may be several such tests, but one is to describe the views in question in coherent prose; or at least, by putting pen to paper. (For all I know suitably gifted people, like Maxwell, could put them in rhyme; as he did with answers to Tripos problems.) [...]

As you surely realize you taking the opportunity of specifying (real) weaknesses in the pdf of 17.VI.07 was—bound to be—reassuring: of (possible) future approval being reliable.

26 June 2007 to John Chancellor. Hitler was a poor student at Linz. (Though born in the same month (April 1889) as Wittgenstein, who was at the same school, he was 2 forms below; especially his German was weak.) He was allowed to move to the next form on condition that he left the school; he went to Steyr. After the *Anschluss*, the SS tried to confiscate all documents in Austria relating to Hitler, but forgot the records at Steyr.

31 August 2007 to Kenneth Derus. *Warning* This is one of my (periodic) letters addressed to you; not because the material is of much interest to you, but because I feel (more) at ease writing about its aspects that attract my attention (more than if I wrote to people eager to 'get on with the job', tacitly, by 'doing sums'). The job is still the formalization of T/W in PA. [...]

In letters to Ridley, I spoke about Crick's antics (which *he* surely did not want to be remembered). When Ridley asked me about my allegedly good influence on Crick I said I recommended him not to say constantly the first thing that came to his mind (and still feel he followed this in molecular biology, discovering his taste and talent for it in *this* area).

Apparently he went to an opposite extreme in areas where he was not among the pioneers. Ridley felt moved to attribute to him *general* cleverness.

7 January 2008 to Kenneth Derus. To-day it is 2 months since my (stupid[25]) fall,[26] and to-morrow I am to move to Fürstenbrunn, where I spent the last 2 summers. Who knows? Perhaps, this change will help to get out of the demoralizing routines of hospitals etc.

12 January 2008 to Kenneth Derus. Kohlenbach's (Herbrand) meeting in September. I don't know if I mentioned any plans around it, for example before my accident. If for no other reason than to distract myself I have tried to think of a title: Aspects of Herbrand's thesis: combinations with contemporary ideas and relations to foundational perennials.

21 January 2008 to Kenneth Derus. *Perennials* include purity of method, but also normal forms (like decimal or binary expansions) and at another extreme (Hilbert's) sums of squares or Nullstellensatz), Herbrand's cut-free deductions from his propositional disjunctions for formulae of predicate logic. Finally, there are *calibrations* by—finite or transfinite ordinal—numbers (cf.: number is 'the' measure of all things). *One* conclusion here is that these perennial idea(l)s *can* be realized, but in the case of (predicate) logic are not rewarding or at least less so than familiar combinations with contemporary ideas. [...]

Disclaimer I have neither taste nor talent for pontificating about pedagogy *pour les jeunes* (a favourite pastime of Bourbaki). In particular, I have no idea of the extent, if any, to which the reminders in the last paragraph might distract (suitably gifted) young people from 'getting on with the(ir) job'.

20 February 2008 to Jussi Ketonen. Admittedly, I was agreeably surprised by your wife's discovery that Wittgenstein's speculations about Longfellow and the Hiawatha rhythm are in print. *Remark* Derus corrected the speculations themselves (in an e-mail to me) pointing out that Longfellow consciously adopted the rhythm of a famous Finnish poem. So it all seems near (your) home.

---

[25] It was stupid inasmuch as I had had plenty of warning that my sense of balance—or, more precisely, my ability to compensate automatically a slight imbalance—was no longer what it used to be. *GK*

[26] On 7 November 2007 Kreisel fractured a hip. He had a mild heart attack eleven days later.

24 February 2008 to Angus Macintyre. Ketonen gives me the (agreeable) feeling of somebody who thinks in general terms effectively (or at least did in the days when he did analytic number theory). – *Remark* His contribution to Goad's project, recently sold to Motorola, in particular, a computer language (SILL) was well rewarded by shares in the business. [...]

What disturbed me was *my failure* to make an elementary proof of Fermat (in his sense of 'elementary') palatable to him.

26 March 2008 to Kenneth Derus. Second letter to you today.

Of course, it is nice to see your list of satisfactory points in H,[27] and that they need not be changed (at least for the moment; it remains to check that they fit the context and, for that matter, the rest of H). But your list of weak passages is more immediately useful: they pick out (for me) where to begin the revision, which seems to be little short of a complete rewrite.

2 April 2008 to Kenneth Derus. Hardy was considered—like Macintyre— to be a successful popularizer; as Littlewood put it he sounded like a missionary preaching to cannibals. (Hardy considered his own style as vulgar; tacitly, intellectually, not socially vulgar.)

19 April 2008 to Angus Macintyre. Now, I hold no brief for the guild of proof theorists; neither w.r.t. mathematics, taste (in elegance), nor even simple literacy (in their way with the vernacular). But not only for you, but for everybody it's hard to judge (reliably) what they would do with knowledge they do *not* have; Friedman is an example of somebody who never took a course in complex variables (on the ground that he *could* learn it if he ever needed it, which Levinson at M.I.T. accepted).

20 April 2008 to Kenneth Derus. I was quite delighted that you picked on Einstein's SO LOGICAL. It was his (Bavarian?) English matching Besicovitch's (he said) Russian English. *Anecdote*. In his lectures he was proud to tell the audience: 40 million Englishmen speak English like you, while 200 million Russians speak it like me (and beamed contentedly). – Obviously, some device is needed—and will probably be found since you pointed out the need—to indicate that a literal quotation is meant.

---

[27]H = Aspects of Herbrand's thèse: combinations with contemporary ideas and relations to foundational perennials (10,452 words, unpublished). 50 complete or partial drafts of this text exist.

What strikes me most about the clumsy passages you list, which will turn out to be a small sample, is how similar they are to the clumsiness in Macintyre's paper (that I don't even mention in my letters to him). Instead of helping me avoid similar (very tiring) tired writing, his example seems to have infected me.

15 May 2008 to Kenneth Derus. It's not just a question of what I should do without you, but of what (my) readers, if any, would do.

18 May 2008 to Kenneth Derus. Of course I was grateful for your help with the PS to [BW] you typed (cf. my letter to Kohlenbach of 17.V.08) and LH. But your help with H has been of a different order; quite apart from your help with BCWW and BSCA,[28] without which I should not have had the peace of mind to write H at all. So, if you agree, instead of concocting an acknowledgment, I suggest you become simply co-author. *Reminder* of other co-authors. Levy, Takeuti, Macintyre contributed in proportion considerably less to their joint papers with me (and I think that by and large both they and I have remained satisfied enough; despite the fact that Macintyre's mathematics in his contribution wasn't too hot, and your help with H has no counterpart to this (negative aspect)).

21 May 2008 to Jussi Ketonen. Nevanlinna / Weil. In his autobiography (*Souvenirs d'apprentissage*) Weil recounts his being condemned to death for being a spy, and being rescued by Nevanlinna, who I suppose sought political asylum in Switzerland after the war. [...]

When I recently—well, about 15 years ago—read *Mein Kampf* again, certainly the style had not become more agreeable; the sentiments weren't either (and by no means original). But somewhere in the middle—of the 2 volume edition—there were a few pages on his prospects in politics. He had just been involved in a failed putsch, he was in jail, with the risk of being kicked out of Germany (not being a German national, and the Austrians rejected responsibility since he had served in the German Army during WWI). So on those pages he listed all the qualities he could think of that have led to success in politics, and concluded he had none of them. His conclusion was that he was a gifted demagogue.

---

[28] BCWW = BlueCard Worldwide, BSCA = Blue Shield of California.

27 May 2008 to Jussi Ketonen. *Reminders* Bieberbach in the same general area as Nevanlinna was an enthusiastic Nazi, too, and Gentzen's unwavering belief led to his death (refusing the help of Polish logicians to have him freed from a Czech internment camp after the war); the Nazi chieftain of Prague University told the professors there shortly before the Russian army occupied the city to stay on, and expressed his regret that he had to leave to attend an important meeting, I think, in Bavaria, a relatively safe part of Germany at the end of WWII.

30 May 2008 to Kenneth Derus. McLarty seems to feel that topoi provided a first occasion when functions are not determined by their values. Ordinary rational functions (= fractions of polynomials) are not determined by the fact that they are both arguments and values $\in \mathbb{Q}$; one uses enrichments (most simply by definitions).

25 August 2008 to Jean-Pierre Serre. As you surely remember, > 50 years ago you described the result—the number of elements of the group $\pi_6(S^3)$, and its description in terms of simplicial approximations—in such a way that, at least for me, my response was a routine corollary from Gödel's proof of (socalled) relative consistency of the axiom of choice etc. w.r.t. other familiar axioms of set theory: It confirmed your own impression; roughly, that the axiom of choice was superfluous, better: that your proof can be specialized to a certain subclass (of socalled constructible sets) keeping the ordinals with their familiar structures, and in particular natural numbers, invariant.

You can judge better than I, how well this result and the general idea are known among mathematicians.

30 August 2008 to Kenneth Derus. In his course (cf. also his book with a preface, which says that the word 'function' does not occur in the text) *Theory of Functions*, Littlewood gave the 2 proofs of the Cantor-Bendixson theorem by Cantor and Bendixson, without and with use of the axiom of choice; more fully, by iterating (transfinitely) the elimination of isolated points on the one hand, and by so to speak beginning by forming the (perfect) set of condensation points on the other. I remember being startled by Littlewood's quip about the latter: They say it uses the axiom of choice; I am sure I don't know. (Wittgenstein proposed the translation 'care' for 'know'.)

2 October 2008 to Kenneth Derus. *Reminder* of a story in Littlewood's *Miscellany* (which he had from Besicovitch) about a Danish businessman known for his impatience with fools. On his deathbed his wife told him he had become kinder. His reply was: weaker, not kinder.

6 October 2008 to Angus Macintyre. As long as I wrote letters of recommendation people were reluctant to express their reservations about me. (Kreiseliana provided an opportunity, but not everyone took it.)

27 November 2008 to Kenneth Derus. Even your comments on Husserl's concerns have not brought them much nearer for me. (I hope you don't feel you have by now been wasting too much time on me.) Of course it is just possible that, if I were in a better state, your exposition would have done the trick.

5 December 2008 to Kenneth Derus. Aristotle's scepticism about—attention to a view of—the middle distance does not apply, as I read it, to a general interest in it; after all, we are constantly confronted by it (and probably pretty well prepared, as they say, by evolution, to use it effectively). But the fact remains that that middle distance looks very diverse, and as Aristotle knew, agreeably so: varietas delectat. The question is, given all this knowledge to start with, how rewarding can one expect closer study of it in terms of ideas about the middle distance to be; compared to relations with micro- and macro-structures. *Remarks* A quite separate matter is the extent to which—feelings of—understanding are evoked by knowing such relations. Speaking for myself or rather those sharing the OR in my BOR polished literary descriptions or paintings of the middle distance may have all sorts of merits for me, but evoke rarely feelings of understanding.

9 January 2009 to Angus Macintyre. My question (which you put to Shelah) is most simply stated in terms of naive set theoretic arithmetic, where well-foundedness is certainly assumed in naive set theory (though rejected as superfluous by Bourbaki because in logical jargon well-founded inner models can be defined in ZF). Find a (non-trivial) naive cardinal Λ s.t., for all $\alpha > \Lambda$, the least in L_α uncountable, strongly inaccessible ordinal of L_α is invariant.

22 January 2009 to Jussi Ketonen. Tarski's excursions into Polish style would-be exact philosophy (or better, formally exact would-be philosophy) aside, which—for all I know—have helped certain students to a career (who had no chance elsewhere), his shift of emphasis *from* provability and decidability (= recursion theory) *to* definability is worth remembering.

2 February 2009 to Adrian Mathias. I remember (what you told me of) your frustrations when trying to convert—the likes of—Mac Lane to share your interests, specifically in set theory. Of course, you are you and I am I, and so would be expected to have different feelings; not only about ideas, but also people. He belly-ached about—what Grothendieck wrote about—axioms of infinity around *étale cohomology*. I told him that, if he wanted to eliminate them, he only needed to specify *suitable* (categories of) functors, so to speak, to provide a thorough algebraization; evidently, a more demanding job. He did not take up this option as far as I know.

16 February 2009 to John Chancellor. Did Dirac speak when we saw him? My feeling was always that he sounded like a voice with a message from another world (as I reported in some article around 1999).

1 April 2009 to Kenneth Derus. I suppose I do not merely feel—in the pedestrian sense of thinking without really knowing—that distraction by idle thoughts is some kind of softening of the brain. But I am reluctant to mention this either to Macintyre or (even) my psychiatrist, who sees in such things a reduction in efficiency to be expected at my age. – By the way, has your experience around Sorabji (who after all lived into his 90s) provided you with relevant information?

14 April 2009 to Angus Macintyre. Perhaps it is an illusion of mine that in better days I'd have spontaneously and immediately thought of the definition by use of AC (and hence in L) of those geometrically pathological 3-surfaces on the sphere that are pairwise congruent and one of them is congruent to the union of the 2 others. – But I had to wait for Derus to remind me of— Mycielski's interest in—them, and the matter of their L-measurability.

12 May 2009 to John Chancellor. Somehow, our last chat lifted me out of my low mood, simply by reminding me of your sister-in-law (whom your father called Miss Freebody, and I only remember as Susie). [...]

Have I ever told you how delighted Verena's eldest daughter was when she learnt that her father was not Verena's first husband, but the musician Paumgartner, a racing car driver and later a big shot conductor and head of the Salzburg Festivals?

6 June 2009 to Kenneth Derus. My first job in the Admiralty was in Mine Design at Havant near Portsmouth. I called Bondi at another department (at Godalming) who had worked on water waves, but had to leave a message: G. Kreisel from Havant called. The message he got was: Christ from Heaven called. Both Crick and I were amused (in those better days), and for a time I signed myself 'Jesus' in letters to him.

18 June 2009 to Kenneth Derus. Your (phenomenological) descriptions would have surely interested Crick, and probably even have been useful to him. Unfortunately, I often had the problem of instability; when I looked longer or changed the distance (or, perhaps, by losing interest in the perception in view); including ordinary optical illusions.

24 June 2009 to Andrew Ranicki. Your father's lecture on Hölderlin reminded me of 2 episodes (in Austria) when I was about 12. Practically all the boys in my form were so-called *Illegale*;[29] probably without ever having heard of Hölderlin they sat around camp fires, and sang of the glories of a hero's death (and with one exception found it in the Caucasus in World War II, for which they had volunteered). Soon afterwards the whole form went to Eisenerz, a Communist miner's village, where we encountered a miners' procession who sang of the glories of dying in the mines (another kind of *force majeure*).

14 July 2009 to Kenneth Derus. Alas, I met Russell only in the mid 1940s, when he was preoccupied with bombing Russia before they had the atom bomb, too, and then (once) in the 1960s at his place in Penrhyndeudraeth in Wales, when apparently remembering the meeting he greeted me with the words that he was consistent. (He did not want the bomb dropped on him; in the 1950s he began to march for peace again.) In those (better) days I found such antics only amusing; as I still find his quip: the 3 volumes of PM are a parenthesis in the refutation of Kant.

---

[29]The Nazi organizations were illegal in Austria after 1934 (when I was 11) until 1938 when Austria was annexed; the Communists were illegal, too, till 1945. Statistically it would be—not only offensive, but—just absurd to speak of *die Gnade der nichtarischen Geburt*. GK

24 July 2009 to Kenneth Derus. In the letter to Levy I can now leaven pedantries about invariance by your information about AZRIEL and AZRAEL. [...]

The fact that Delzell typed his dissertation without any typos with 2 hours (transcendental) meditation a day and 2 hours sleep a night is unimportant mathematically. But perhaps it is a counterpart to Beeson who had imaginative ideas while taking LSD, when he couldn't type a single line properly, let alone, put the ideas into the form of a proof (until he was off the drug).

30 July 2009 to Hubert Faure. In better days (I hope) I'd have been more alert than this morning when you called about van Heijenoort, specifically about his having been murdered by his fifth (and last) wife. Presumably this was not said about her in the book (about Trotsky and Stalin). There was a real estate boom in Mexico City in the 1970s from which she had profited (but did not get out of the business while the going was good).

Presumably I never told you—and certainly don't remember telling you—that van Heijenoort studied mathematics in France, then joined Trotsky for 7 years (but did not speak about him when we visited the Romanovs at Woodside outside Stanford). After Trotsky's murder, he moved to NY (and joined the Trotskyist circle around Mary McCarthy, decidedly more articulate than v.H.[30]). He attributed his (successful) career move to the history of logic to Trotsky's practice, whose speeches were full of history, usually beginning with the Romans, but adapted to his own very different temperament.

4 October 2009 to Kenneth Derus. Am I just imagining that your reference to Wittgenstein's inept adjective *bunt* (and Miss Anscombe's even more inept translation) has come up in our correspondence before?

20 January 2010 to Juliette Floyd. Wittgenstein once told me that there were some things that would kill him, but not me, and many more that would kill me, but not him. In plain English I take this as a warning that it would be unrewarding for me to speculate about his philosophical views such as those you mention in your (short) letter.

---

[30]It's quite likely that I told you about dinner with her at James Jones' flat on the Ile St. Louis where both he and she were pleased when I noted that, unlike him, she wasn't a real novelist; he couldn't hold any other job (if he tried), in contrast to her who'd be a success at any she tried. He recalled a long list he tried and at which he failed. *GK*

20 January 2010 to Kenneth Derus. Ann Getty, one of Paul Getty's daughters-in-law and (former?) wife of Gordon Getty with ambitions to be a composer once asked me if I had ever met an intelligent banker. In her presence her husband never said a word (but was quite talkative without her). My impression was that he was not at all afraid of her, but just proud of her. – After an evening with him I gave him a reprint of my obituary of Russell because I quoted his father's autobiography in it. (The son thought his father would have been 'tickled pink' if he had known. So might have been the Hungarian (?) psychologist who thought that diverse reactions to paradoxes were reliable indicators of different mental abnormalities.)

22 January 2010 to Kenneth Derus. Broad's style, also in chats when we met in Great Court at Trinity, was not merely fluent, but polished. [...]

(d) Subject to your correction, I continue to believe that I remember 2 out of 3 items in the autobiographical note which he had always wanted, but never dared to do: enter a night club, and stroke the hairs (on the head) of a Swedish boy. When von Wright first came to Cambridge in the 1930s, he must have looked pretty young; he was certainly fair-haired, but though a Finnish national, he was of Scot-Swedish descent (and as Ketonen will tell you, paid the highest taxes of any private Finnish taxpayer; presumably firms like Nokia paid more taxes at the time).

26 January 2010 to Kenneth Derus. *Anecdote* about Wittgenstein. He told me I was green and that mathematics kept me green. (He said this when I giggled in his socalled seminars, in effect, as an explanation why it was OK for me to do so, but not for others.)

27 January 2010 to Angus Macintyre. This letter is prompted by an observation of Derus on the difference in style between my letters to you and to others; tacitly (I think) other letters in English (perhaps not to Kohlenbach; though Derus knows German he is perhaps not comparably sensitive to style). I have not paid attention to style, but certainly I am aware of a difference in my feeling when writing to others: a sense of futility. [...]

Here (with my teenage experience of analytic determinacy in hydrodynamics, but not continuous determinacy) I perhaps inevitably find Köllner / Woodin's assumption of bifurcation between $2^{\aleph_0} = \aleph_1$ or $2^{\aleph_0} = \aleph_2$ simply short-sighted. Admittedly, not back in the 1940s, but over the decades I learnt to correct my (mathematical) teaching at Cambridge, where Hadamard's principle on continuity in parameters was presented as some kind of ideal for physical relevance. *Discovered* extremes among alternatives are, at one extreme, some kind of smoothing of the results (suppressing high frequencies in Fourier expansions) and, at another extreme, topological properties such as different kinds of attractors (of Thom, Zeman, and the Russians).

29 January 2010 to Kenneth Derus. Broad. Perhaps it is superfluous to repeat that I have often published (though perhaps not written to you) about my distaste for—what struck me as—Gödel's would-be flashy and Wittgenstein's affected writing and its sharp contrast to my conversations with them, especially during the first few years after meeting them. (Who knows to what extent their final illnesses or just the wear and tear of years of acquaintance reduced my—and for all I know their—interest?) I don't know if I'd use the word 'like' about Broad with whom I chatted long before I read anything by him, but I certainly have agreeable memories. For example, several times when we met he quoted verses in German (in a perfectly competent accent). I asked him if he was sure about the 4th (or 6th or whatever) line, and he was not offended (though I am pretty sure he realized I had no doubt).

5 February 2010 to Hubert Faure. Did Iris Murdoch really make copies of her letters? (Those she wrote me were as hard to decypher as Chancellor's, but when decyphered not comparably as stylish as his.)

7 February 2010 to Angus Macintyre. *Anecdote* My conversations in Jan 1942 with Wittgenstein about Hardy's Pure Mathematics, which presents analysis from a single axiom (the principle of a least upper bound for any bounded set of points). I just reworked Hardy's proofs in a way that was familiar to me and felt suitable for Wittgenstein, who said in Cambridge fashion of the day that he did not 'understand' them (and he seemed satisfied). [...]

Actually, Wittgenstein's own completeness proof for propositional logic by truth tables in Tractatus, written during the war 1914–18, and published at about the same time as Post's, was of course never 'discussed'. It had been overlooked in the literature. One shrugged one's shoulders, and said, the *derivations were not complete for proofs.* [...]

Anecdote Girard had the habit of attributing to me classical quotations which I had happened to mention to him. I suppose from his severely self centered view this is consistent.

16 February 2010 to Kenneth Derus. For all I know Wittgenstein liked making up metaphors, but wasn't much good at it. If popularity is used as measure of aphorisms his seem pretty good, though (for me) not comparable to his theatrical gifts in everyday life. Have I not told you of my discovery after his death that—when we regularly went for walks after lunch, and he quoted aphorisms—he had noted them down according to his diaries a few days earlier.

20 February 2010 to Kenneth Derus. Wittgenstein's perfectly sensible prose about nonstandard models and equivalences between Π_1^0 sentences and insolubility of diophantine equations is—surely less vulgar than Turing's test or Gödel's Gibbs lecture, but—decidedly less demanding (for example, in BOR) than the discovery of effective uses. After all, Abraham Robinson's uses of non-standard models were not enough even for an FRS.

23 March 2010 to Kenneth Derus. Macintyre told me (last Sunday) that the category theorist Lawvere with foundational interests was Truesdell's student (later presumably associated with Mac Lane). I think I met him in the 1950s, actually before I realized the potential of enrichments, let alone, their logical aspects in functional interpretations. (Actually, in lectures at Stanford that Tait wanted to preserve for eternity, I systematically used enrichments before knowing a word for it.) [...]

I don't really know what to say to Parikh, who has told me for the fifth time (also in person) that he remains close to his divorced wife.

28 April 2010 to Kenneth Derus. I haven't really attempted, let alone, succeeded in absorbing the inwardness of Badiou's text.[31]

18 June 2010 to John Chancellor. Derus has drawn my attention to my many sloppy formulations in recent letters (showing at the same time that he understands my meaning by supplying elegant reformulations). Please, let me know when this becomes intolerable to you. *Afterthought* Please, let me know too, if and when you are able to print out (and read my) scanned letters. Apparently, by now Derus has stored > 4444 pages of correspondence (by and

---

[31] To me it sounds like juvenile effervescence which I find or, at least found, charming, for example in Hofstadter when I met him at 14 (his father, the physicist, was professor at Stanford), but was bored by Gödel, Escher, Bach, which won a Pulitzer Prize. *GK*

to me) in his computer, and so it is time to remember that there are points of diminishing returns.

29 June 2010 to Angus Macintyre. In the 1970s, Jensen spent some time at Stanford, and lived with Mathias at the house I rented at Los Altos Hills. (I am sure I told you how Jensen set the house on fire twice trying to force a 3 pronged plug into a 2 hole socket (or the other way around), presumably evidence for—among other virtues—persistence.)

19 July 2010 to Kenneth Derus. As you'll see in a moment several items in your e-mail of 19.VII.10 are of immediate use to me (without having properly taken in the rest). Most of all, w.r.t. Mints, who restarted the correspondence after many years, reminded me of a conversation with Wittgenstein which I have probably told you before. He cited the French proverb: On ne peut pas chier que son cul. I replied that—as I had actually been told by a professional nurse at Addenbrooke's—some try, but admittedly make a mess. (As I have probably also mentioned, he was happy to make would-be daring quips, but less so when I continued in—what I felt—a similar vein.)

6 September 2010 to Kenneth Derus. PS about Kripke. Have I ever told you the anecdote about him when he spent a summer at Stanford, and was not let on a flight back to the East Coast because his 'hopping about' was interpreted as evidence for his being on drugs?

13 October 2010 to Kenneth Derus. Ich habe einen Nackenwirbel angebrochen und bin im Spital: DIAKONISSEN SALZBURG. Bitte informieren Sie Blue Shields.[32]

22 December 2010 to Kenneth Derus. Natürlich ist das in meinem letzten Brief Ihnen unterstellte Ziel, zu tun, als ob alles beim Alten geblieben wäre, nur eine Vermutung, aber ich hoffe sehr, daß Sie der Versicherung nicht eingeredet haben, daß es sich um eine folgenlose Verletzung handelt.[33]

---

[32] *I have broken a vertebra in my neck and am in this hospital: DIAKONISSEN Salzburg. Please inform Blue Shield.*
On 9 October 2010 Kreisel fainted, fell down some stairs, and fractured the body of C2 (not the dens process). This was repaired with screws. Access to the vertebra was by way of his throat.

[33] *As I said in my last letter, you appear to want to act as if nothing has changed. Admittedly, this is just a guess, but I very much hope that you have not persuaded the insurance company*

20 April 2011 to Kenneth Derus. *Remark.* Husserl's phrase, which you translate as 'non-calculating glance', presumably something like 'unbefangen Blick' assumes that socalled immediate perception—without theoretical prejudice—gives access to an essence; cf. also the idea(l) of authentic representation.

23 April 2011 to Kenneth Derus. 2. In proof theory, equivalences paid attention to in recursion theory are not good enough (let alone ordinals of the order types in view). So reviewed the countability (in V) of $\aleph_{USI}^L$ is neither here nor there where it is to be added that the ordinals satisfying a definition for $\aleph_{USI}^L$ in different segments L_α (satisfying, say, ZF) change up to a certain limit. So the symbol $\aleph_{USI}^L$ denotes different ordinals, but again only below a suitable bound, hence, asymptotically stable. *Remarks* This property (of asymptotic stability) of operations on infinite ordinals is a counterpart to closure of finitary operations on finite ordinals. Whether statements of this fact in mellifluous or majestic prose are comparably memorable to extended practice with short comments is, at least to me, open.

3. *Re* non-calculating glance and authentic representation, 2 expressions you have used (repeatedly), and I mentioned recently. (a) To put first things first, they are both clear inasmuch as German equivalents strike me immediately (voreingenommener Blick, authentische Repräsentation) with definite feelings, albeit oversights on second thoughts related to illusions of uniqueness, equivalence and the like taken literally overleaf. *Remark* They are catchy phrases, at least if I have quoted them correctly. I do not know if Wittgenstein would have called them bewitching. For many of us they are the opposite: the moment they are mentioned they draw the attention at least of some of us to widespread oversights even if one hadn't noticed them before.

15 May 2011 to Angus Macintyre. My impression of being more and more slow-witted may be a little exaggerated by socalled clinical depression, but it is bound to be intolerable for normal people, if it isn't already. *Reminder* In his *Miscellany* Littlewood tells a story of some Fellow saying to Miss Cartwright: 'I may be stupid'. Her reply: 'Yes, I think so'. (I got on very well with her. Did you ever meet her?)

10 September 2011 to Kenneth Derus. As to your #9, taken literally it states that, except in dimension 4, the exact number of non-equivalent dif-

---

that my injury was inconsequential.

Chitchat with the Devil 157

ferentiable structures on spheres of all dimensions ≤ 20 is known. [...]

Are you not put off by the sheer triviality of all these, let us say, asides compared to your #9 (far beyond my wildest expectations)?

15 June 2011 to Angus Macintyre. Even somebody dripping with affectations like André Weil or at another extreme like Thom, who pretended total ignorance of physics and did not believe in atoms resp., chose affectations that did not interfere with their proper work.

17 June 2011 to Henk Barendregt. Your mother asked me how I broke my neck; more precisely, I damaged the first 2 vertebrae, which were screwed together. Actually, I have no memory of it at all; the doctors speculate I may have been unconscious before I fell backwards some 4 or 5 steps on a familiar staircase. *Remark*. Before that accident I was in reasonably good shape, and spent my days as your mother described hers in her e-mail, taking daily walks for exercise.

25 June 2011 to Kenneth Derus. In my early teens I found the science and mathematics I encountered at school boring compared with German literature and even Latin.

29 June 2011 to Kenneth Derus. Have you heard of the Swiss institution DIGNITAS?

5 July 2011 to Henk Barendregt. Of course I hold no brief for Dennett's ideas. But I know he was a chum of my late friend Crick (for whose ideas I—have learnt to—hold no brief either).

10 July 2011 to Kenneth Derus. I just received a request for a contribution around van Heijenoort. If I may I'll ask you for references in the Templeton volume to A. Weil or I. Berlin for a piece with the provisional title Memories of some uses of (apparently) scholarly history of ideas. With luck this will not be another stillborn project.

14 July 2011 to Kenneth Derus. Pompidou was director of Rothschild in France before de Gaulle persuaded him to go into politics. He knew Faure very well. Over dinner Pompidou told us the Rothschilds had a slogan around the question of losing money: the quickest way is gambling, the most agreeable is interest in women, the surest (in the 19*th* century was) speculation in railways shares. In the 20*th* century, the first 2 were the same, but the surest was following the advice of consultants.

22 July 2011 to Kenneth Derus. Chancellor raves about your letters. I'd have thought he would send copies of his e-mail to me also to you. [...]

As to Barendregt himself there is no doubt that he quotes correctly what I said to him when I lived at his mill in, I think, 1971, but none of it seemed memorable at the time (or now), in contrast to various things I had said to him, and probably even published around that time. (Your selections from my letters in our correspondence strike me as much closer to my own emphasis.) [...]

You recently wrote that nowadays our letters are memos to ourselves, tacitly without attention to style. To me there is an asymmetry. Mine are really undisciplined (and only the exceptions you choose for comments save them).

21 August 2011 to Kenneth Derus. 1. First of all, I remember you told me repeatedly you liked my literary style (tacitly, in my better days). Quite simply both D7 and D6 remain desperately clumsy (even after your corrections).[34]
2. You explicitly stressed that your interest was *not* primarily in the logical aspects. But it should not be forgotten that, albeit in other connections, you also stressed your interest in—what is sometimes called raw experience, in good company (of Goethe, who also liked writing endlessly about and around it). At least in the van Heijenoort article, especially sections 2 and 3, there is a *principal theme*: technological auxiliaries (to natural history of raw experience).

8 September 2011 to Kenneth Derus. Of course van Heijenoort's interests in refutation trees were complementary to mine. He crossed all the *t*s and dotted the *i*s in the proof theory—not only of classical, but also—of intuitionistic and modal predicate logic, while I viewed (classical) refutation trees as codings of *all* (countable) countermodels of the end formulas related to the non-counterexample interpretation; in short, squarely in model theory. [...]

Certainly your idea of my (sudden) faints fits all the facts of which I am aware.[35]

---

[34]D = Van Heijenoort's taste for (suitable) attention to (suitable) vagaries (9,059 words, unpublished). 25 complete or partial drafts of this text exist.

[35]Kreisel fainted and fell down a few stairs on 1 September 2011. He was hospitalized but

13 October 2011 to Kenneth Derus. Instead of describing my state since arriving at Oaken Holt,[36] where I don't sleep and have an unreliable doctor, I can refer to the trial of Dr. Conrad Murray, and the state of Michael Jackson the day he died. It is very likely he felt the way I am feeling now.

24 October 2011 to Kenneth Derus. It does not seem an exaggeration if I say I am afraid of losing my mind.

16 December 2011 to Kenneth Derus. Abstract genetics of the first half of the 20th century introduced the idea of genes in contrast to molecular genetics (and epigenetics) of the second half. Do you know a parallel in the first half to the (of course equally abstract) neural nets of McCulloch & Pitts in the 1940s and 1950s; with Kleene's additional relations to finite automata?

5 January 2012 to Kenneth Derus. While Russell was exceptionally urbane, Zermelo seems to have been clumsy; for example, (equally clumsy) solemn Swiss logicians were indignant over Zermelo's note in some (Swiss) hotel register under the heading *nationality*: Gott sei dank, kein Schweizer.

6 January 2012 to Angus Macintyre. In the turgid obituary of the *Times* Dummett is called irascible, followed by the familiar pious psychobabble on his high minded motives. *Aside* Perhaps many people are (irascible); it just so happens that this reaction is alien to me. For all I know Derus is a bit impulsive, too. Occasionally, I refer (obliquely) to an outburst of his in the more or less distant past. His reply is almost invariably a quotation from a quite recent letter that does not fit the (past) event at all. As the Beatles used to sing, I 'let it be'.

10 January 2012 to Kenneth Derus. Admittedly, I speak slowly compared, for example, to Fr. Heinisch. But literally every time that people at the hotel try to complete a sentence for me they say the opposite of what I (consciously) intended.

16 January 2012 to Kenneth Derus. Years ago in some letter (also to you) I used the phrase 'chitchat with the devil' for Goethe's use of the—in fact,

---

not seriously injured.

[36] Kreisel traveled to England on 1 October. Throughout his visit he suffered from intractable insomnia. Benzodiazepine withdrawal was a factor.

almost totally unrelated, but—familiar tradition around Faust to give him a literary frame to talk about everything under the sun. As I see it van Heijenoort's quotation about vagaries gave me a literary frame to talk—not about as many topics as in *Faust*, but—about a lot (almost totally unrelated to van Heijenoort's interests).

17 January 2012 to Angus Macintyre. Littlewood did not look distinguished at all but Hardy certainly did, and so did Adrian, the neurologist. (Littlewood looked a bit like a well-preserved boxer, and in my memory no less distinguished than, say, Mohammad Ali.)

24 January 2012 to Hubert Faure. I must have told you that one of Iris Murdoch's (few) memorable remarks to me (about me) was that I lacked any sense of the urgency of action.

2 February 2012 to Angus Macintyre. Nobody I know who knew Lighthill believes that Lighthill committed suicide in proper Roman fashion when he died swimming round an island nor that he was unaware of the danger that such an exercise presented at Lighthill's age. [...]
 Neither Dyson nor I ever had strained feelings about the other. He realized that his first wife was a bad choice for him, she had found a maid for the family, whom he married, and seems to have lived happily with her up to this day. He was, and probably still is, exceptionally gifted mathematically, but did not find it satisfactory. *Anecdote* Friedman's first wife came to see me after she had prepared lunch for her husband, Shelah and some others. I remember her words: Guess what! Those guys really believe everybody would want to be a mathematician if he could. (I wanted to mention Dyson as an exception, but she wasn't interested.)

3 February 2012 to Kenneth Derus. Forgive me, I am exhausted by now 3 weeks of constant diarrhea. (The doctors have not found a cause.) Writing Section 5 of the v.H. piece is my only relief from that weakness and from most all of the nurses. Should I stop sending you the stuff?

9 February 2012 to Angus Macintyre. Lighthill complained quite openly about the fact that in their school days at Winchester Dyson was always 2 steps ahead of him; about the fact, not any injustice. During the war matters changed. Dyson did operational research for the RAF without any taste for it (and distaste for Air Chief Marshal Harris), while Lighthill did research on aerodynamics and hydrodynamics in a department led by an FRS who recognized Lighthill's gifts.

Though Dyson has decided on a literary career (and I find his literary style only agreeable, but less impressive than his mathematics) his literary style infuriates his first wife.

Kant's literary style was called in England not bad, but mad (at least in the 1940s, if not the 1960s). This certainly applies to the English translations, but not to the original, which a 12 year old can read without difficulty.

12 February 2012 to Angus Macintyre. In the 1940s at Trinity I joked—admittedly, in very bad taste—that, for me, Hitler was not an unmitigated curse since otherwise I'd not have found myself at Trinity, and met Littlewood and Dyson (as a teenage friend).

14 February 2012 to Angus Macintyre. Derus seems to have a great liking for using strong language and I to provoke it (in Leibniz's best of all possible worlds).

19 February 2012 to Angus Macintyre. Bernard Williams told Hubert that he thought I did not like him. Well, my impression was that *he* was ill at ease in my presence, in contrast for example to Ayer, who also struck me as too light weight for words, but, at least for me was agreeable company. So was Strawson.

21 February 2012 to Angus Macintyre. Dyson's war work on operational research did not satisfy him, nor apparently his proof of the equivalence of Feynman's, Schwinger's and a Japanese's formulations of quantum electrodynamics. *Anecdote* recounted by Dyson (about Feynman). When Dyson presented the equivalence at some meeting of Big Shots, Feynman said to him: 'Doc, you are In'. Feynman's language is just as light and airy as of those black queens whom I met with Montague.

23 February 2012 to Angus Macintyre. At school we learnt about King Charles (II): He never said a foolish thing, and never did a wise one. Of course this is hyperbole. I wonder if the same applies to me. Of course there are businessmen like Suppes who never did a foolish thing and never made an interesting observation.

IAS. Gödel's wife called it a home for old age pensioners in the mid 1950s when he was around 50. The emeriti I met then were Morse and Veblen, both (to me) in their different ways—what I imagine to have been—Americans of the old school. (Veblen was surely a New Englander, Morse may have been from the Midwest.) Compared to them people like Weil (though not Harish-Chandra) struck me as clowns. [...]

27 February 2012 to Angus Macintyre. Once again Derus. Actually, at least as I see him, what strikes you as blunt in ordinary conduct, seems to me related to his very decisive views on subjective experience. (Fortunately for all concerned I heard those views when I was not exhausted by weeks of diarrhea, apparently well known to be tiring.) [...]

I am sure I have told you of one of my encounters with Brouwer when I had noted that he mumbled a classical proof in Dutch, only to translate it into an intuitionistic proof in his lecture (in a sort of English). When I mentioned this he quoted George Bernard Shaw (as if one hadn't heard this a hundred times): One must exaggerate to make an impression. When I reminded him that G.B.S. did not promise him that he'd make a good impression, he went off without saying a word, aka in a huff.

4 July 2012 to Angus Macintyre. As you surely remember Einstein liked to pronounce misanthropic opinions, for example, about his pride in surviving Nazi domination of Germany, which was exceeded by surviving 2 marriages. I too survived Nazi domination, and I didn't even try to survive marriages. But I see nothing to be proud of either. Admittedly, I sometimes feel satisfaction to have survived doctors and other medical attention, but not for long: pride doesn't seem to loom large in my vocabulary.

30 July 2012 to Kenneth Derus. Your quotations from my well-wishers[37] remind me of a Mitford sister's malaise over the clumsy attempts of—tacitly, would-be well-bred—Americans to write polite English.

10 August 2012 to Angus Macintyre. *Reminder* of Tao's success with—the rediscovery of—the no-counterexample interpretation which he renamed 'metastability', and used relatively memorably.

---

[37]*Afterthought* If they think I am about to die, it is time to apply *de mortuis nil nisi bonum*. Roman historians did not often follow this principle. *GK*

31 August 2012 to Kenneth Derus. Certainly, Wiles' proof would have been beyond Fermat, but not Gentzen's ordinal inductions up to ω^ω etc. [...]

In better days I was interested in the hunch (of several people including myself and the hydro-dynamicist Lighthill) that Fermat 'proved' his remark by using an ordering that *looked* like a well-ordering, perhaps to be found from a proof-theoretic analysis of a proof in EFA.

3 September 2012 to John Chancellor. If I have the energy I'll write to Nedo to ask if he knows more about the (literary) style of other purchased Aryanization documents (in accordance with Göring's public dogma: Wer Jude ist, bestimme ich). *Anecdote* Wittgenstein told me that in the course of this transaction he met Göring, and found him agreeable; in fact, he felt sorry for Göring when he read of conditions in Göring's cell after being sentenced to death in Nuremberg. As far as I remember Wittgenstein paid no attention to Göring's drug taking.

8 September 2012 to Kenneth Derus. Wittgenstein was sceptical about Paul Wittgenstein's musical gifts, especially by comparison with another brother who had committed suicide at an early age. [...]

During Wittgenstein's lifetime I had no idea that he worried about his lack of originality, as if there was only one kind of originality. But soon after his death his list of aphorisms (Value and Culture) was published where his particular kind of lack of originality was attributed to his not being of Aryan descent (despite his Aryanized grandfather). Whatever one may think of such reasons there is no doubt that taste in expositions differs widely. – *Anecdote* Wittgenstein was reported to have said that he did not mind what he ate, as long as it was the same every day. At the time I ignored it as a conversational gambit (and I'd continue to do so if I had not the impression that you like to collect such tales). *Remark* I also have the impression that you have got used to lively letters before my decline during my last stay at the Diakonissen.

21 October 2012 to Barry Mazur. Perhaps you remember our chats at Bures in the 1970s or 1980s about André Weil's (understandable) helplessness around the general claims about Gentzen's work in the 1930s as proving the consistency of socalled Peano, aka first order, arithmetic (PA) by means of, tacitly, logic-free ϵ_0-induction. [...]

(Angus) Macintyre has lectured on the possibility of formalizing Wiles' proof of Fermat('s marginal remark) in PA, in particular, on the arithmetization of Wiles' uses of étale cohomologies. Whether or how much further detail is needed to apply Gentzen's logical work depends of course on what is wanted, for example, by historians of mathematics. I don't know any. Do you?

Impressions. Presumably, the actual ordering along which the descent takes place and the predicates to which the ordinal induction applies—tacitly, defined in contemporary terms—are of primary number theoretic interest, rather than the ordinals. *Anecdote*. When I first learnt Gentzen's proof I (frivolously) hoped that a modification—of Gentzen's transformation—would suggest itself with a simple ordering that was not well-founded, and thus a truly miraculous illusion of a proof.

29 December 2012 to Kenneth Derus. I looked in a mirror today while shaving. What I saw was strange, but more dead than alive; admittedly, some people die at 60.

21 January 2013 to Angus Macintyre. *Warning* for people (like myself) bored by actual cut elimination. Even the bounds in terms of (interlocked) recursion are only *upper* bounds. [...]

Anecdote Until about a year ago I found writing much easier than explaining proofs in conversation, when I was easily impatient. Now it's the opposite.

29 January 2013 to Kenneth Derus. Fr. Heinisch is just 70, but so undisciplined that she starts a sentence, and then doesn't remember the name of the person in view. Specifically, yesterday one of her sons who is in the music business apparently is very much impressed by your book about X. I asked if X was Sorabji. She immediately said 'yes', but so unconvincingly that I regretted having given the first name that came to mind.

21 February 2013 to Kenneth Derus. Nedo sent me 2 books; one by Sebald, Austerlitz (about 400 pages); one by A.V. Thelen, Die Insel des 2. Gesichts (Mallorca; > 900 pages). Both are largely about emigrants (like Kissinger or myself). Neither comes close to Kissinger's aperçu: Emigration was felt as a catastrophe by people of his parents' age, but for—tacitly, the likes of—him, it was no more than a blip, and in his particular case a Foreign Secretary of the USA, especially during a war, counts for more than a professorship at Göttingen or Heidelberg.

1 March 2013 to Angus Macintyre. Derus has cleared up enough (at least for me) the terminology 'modularity property' of elliptic equations; certainly enough for me not to try to *chier plus haut que mon cul.*

11 March 2013 to Kenneth Derus. In the 1950s my preference for Littlewood having written Skewes II was not some kind of eccentricity on my part; among other things it showed that Littlewood understood Hilbert's *Ansatz* in proof theory well enough to apply it in number theory. I suppose that nowadays it would be considered some kind of preciousness at best.

22 April 2013 to Kenneth Derus. By the way at the time I introduced the colloquial expression of unwinding, I most certainly considered the extraction of a bound from proofs of $\Pi_1^0 \to \exists_1^0$ and $\neg\Pi_1^0 \to \exists_1^0$ as obvious. When I first learnt Littlewood's proof in lectures in my teens I was told that this was not considered obvious.

31 August 2013 to Angus Macintyre. My first use of proof theory in JSL 1952 is not well known; I think even Kohlenbach isn't happy with it (in contrast to the proof-theoretic analysis of Artin's solution of Hilbert's 17th problem). *Anecdote* Artin had spent his last semester in Germany on bounds for his solution, which I had not known in 1955, when Gödel told me of that open question, and sent me to Artin. In no longer than one of our telephone chats I explained it to Artin, who summarized it painlessly (to him and to me). (a) A polynomial with coefficients in an ordered field and ≥ 0 in its real closure can be written as a sum of squares, and (b) the adjunctions can be eliminated (in the Galois tradition, which you, but not Artin emphasized). All this would have been understood by members of Artin's seminar. To complete this broad idea quantifier elimination, known since the 1920s, was enough. – I myself found this convenient for explaining the potential of proof theory around logic generally, but not demanding. (Kohlenbach seems to like it, too.)

2 September 2013 to Angus Macintyre. Whatever my personal interest, perhaps Sarnak would be interested in Skewes (1955). Though Littlewood had told Besicovitch that he had written the paper himself, a moment's thought is enough to see that (a) Littlewood had provided the many ideas for which Skewes thanked him anyway, but (b) could not have written the footnote on p. 67 even when drunk (almost every day after dinner).

22 September 2013 to Dana Scott. My main complaint about George Dyson's book was that though fluent it was not as amusing (to me) as he was at the age of 5 or 6. *Anecdote* He had spent a summer (with his now famous sister) at Stanford, and I drove into a telephone pole. The following year I met him at Princeton, and I mentioned the mishap. He reassured me: At least something happened.

30 October 2013 to Kenneth Derus. Gödel's English was better than Girard's. But Gödel was not too vain to have his texts corrected by Church. Now, I don't assume that Girard would be able to find anybody of Church's pedantic skills to be prepared to correct his clumsy prose. But with a little more self-confidence Girard would publish in French. *Disclaimer* I hold no brief for Girard's thoughts. But their formulation in French couldn't be as clumsy as in English (which I don't find in the least entertaining, but only tiring).

31 October 2013 to Angus Macintyre. I feel I am declining rapidly, and want to warn you to be realistic; not to talk about it at any length, let alone, solemnly, but not to forget it.

13 December 2013 to Kenneth Derus. By the way Ingham 1935 anticipated almost all ideas that Kaczorowski impressed me with such as the almost periodicity of $\pi(x) - li(x)$. (Ingham knew this idea from H. Bohr's 1925 paper 10 years earlier in Acta Mathematica.)

26 February 2014 to Angus Macintyre. I remember the difficulty I had to find a topic of conversation (or correspondence) when Bernays was dying. [...]
   Admittedly, I may recover, and we may have chats as in recent years: But it would be as unrealistic for us to dream about that as to instruct Sarnak.

20 March 2014 to Kenneth Derus. My medical knowledge is very restricted: I don't even know what is common knowledge either among professionals or lay people. For example, at least in Austria pneumonia is apparently common after operations; in my case on both sides of the lungs with water collecting in the lungs (1 litre on one side, 800 ml on the other). This was tiring till most of the water was surgically removed.[38]

---

[38] The pleural effusion was a sign that Kreisel's heart was failing. (The water was on his lungs, not in them.)

29 March 2014 to Kenneth Derus. I showed your letter to the doctor attached to the Seniorenheim (Nonntal). He agreed with your diagnosis, either because he understood (the English text) or because he didn't. The doctors here are too vain to admit ignorance of English.

The doctor also corrected my ignorance. When I was asked if I agreed to have the water in my lungs sucked out, I thought that was a routine question before any operation. According to the doctor it was not, but came under the heading of life-saving interventions. Any patient (here) has the right to let his life end in a natural way (with palliative medicines to avoid pain). I really don't know what I'd have said if I had understood this at the time; cf. PS. [...]

PS Not even, but especially, on second thoughts I am sure I'd have been disturbed if I had understood the question. Fragments of thoughts would have come to mind without any hint of the knowledge needed for an answer. The surgeon left it to me to decide between a general and a local anaesthetic. I chose a general and was elated when I woke up (feeling inebriated). Did I have enough knowledge for a decision, for example, about the state of my heart? *Anecdote* Teenagers in England—but not I—when I was in my teens talked a lot about suicide, but I remember only 2 who actually did (and none who had *not* talked about it before killing themselves).

If I am not mistaken, what prevents—at least the likes of—me from talking about my death is simple embarrassment at dramatizing one's condition. It was quite different when the doctor brought up the topic. Alas, he had little that was at all memorable to say. His companion, a woman doctor, assumed that I was unquestionably glad to be alive.

17 May 2014 to Kenneth Derus. The urologists here—not only Schmeller who recommended my most recent operation after which I had water on the lungs—say that the condition of my heart was the cardiologist's business not the surgeon's.

25 May 2014 to Angus Macintyre. At least in my own case I am unable to pursue non-mathematical interests for any length of time (much less than in better days).

10 June 2014 to Angus Macintyre. *Warnng* I found Littlewood (at my present age) charming, and believed (therefore?) that he understood what I said about mathematics. Now, I know that he did not or else did not pay attention to what his pupil Skewes did with it.

1 January 2015 to Hubert Faure. Thank you for letting me know about John Chancellor. His last e-mail to me was dated 14.XII.14 (and was about the funeral in Cambridge of the widow of a man I knew, too, and on whom Chancellor had an intellectual crush). Since then we have chatted on the phone (at Xmas).

Presumably, you will be meeting his children soon. Please tell them I am sorry not to be able to attend his funeral. Actually, I have known him longer than the children have (known him). We first met in 1946 at dinner at Trinity. He had sympathy with my foibles such as my inability to attend funerals.

9 January 2015 to Kenneth Derus. Faure just phoned, telling me that Chancellor's family wanted from me a text that can be read out at his memorial service. Since you are one of the few people who can still read my handwriting, I wonder if I may send you a text (to type) next week. The service is on the 22nd (this month). [...]

Have you ever heard—the locally common tale—of a weak heart being responsible for a constant feeling of tiredness? pedantically, perhaps not constant, but starting at about 2 hours after breakfast, and then lasting all day? And is there a remedy (known to you)? perhaps at the price of shortening one's life. (This would be cheap given the value of that tired life.)

15 January 2015 to Kenneth Derus. Please, arrange with Hubert Faure, where to send the typescript below for Chancellor's Memorial Service.

Of course, this text is around my friendship with J.P. or J.C. (short for John Paget or John Chancellor), which lasted to his death. We met at dinner at Trinity in autumn 1946 when he was still in his teens (and I was a little older). We rarely talked of solemn matters, but one day he brought up the topic, what was special about our friendship. He startled me with the news that (a) everybody knew how easily he was hurt by criticism and that (b) I had never done or said anything to him to hurt him. In the spirit of the day or perhaps of the society in which we moved I (truly) replied that I had never noticed his fragility (though, because of his appearance in his youth he was sometimes called Dresden China).

If I had had the sense to write an obituary years ago, as professionals apparently do, it would surely have been more polished and agreeable to read than this weak composition. But perhaps there is a compensation since I have now something to report around recent exchanges with J.C.

(a) Our last chat on the telephone was on Xmas day 2014. (We had kept in touch ever since we first met.) I had noted nothing exceptional, though

(b) on 22.12.14, after some e-mail correspondence with me he had written Dr. Derus, who typed my e-mails (since J.C. had not been able for a long time to read my handwriting), that he was in a worse state than I. (I saw this only in 2015.) In short, he had a better understanding of our state than I.

(c) Of course, I have long known the phrase about people being 'stunned' by the death of a friend. But I at least was totally unprepared for my own experience. After I had heard of his death, I went to bed, kept all night long awake, stared at the ceiling without remembering a thought that may have crossed my mind, and remained—still wordlessly—sad in the morning. In short, I had been stunned without having previously guessed at that meaning.

To return to the stability of our friendship, to which I gave no answer when he brought it up; of course this is easier to judge now than some 10 or 20 years ago. As I see it now it comes under the heading: Accentuate the positive, provided you are by temperament able to do so. (In the case of J.C. I did so effortlessly.) I can well believe that not everybody did so effortlessly.

What is more in all the conversations I remember he offered something substantial to think about, even in our first conversation which I remember only partially. It was a piece of wisdom he had learnt from a schoolmaster, which in itself was unusual among undergraduates.

By the way, many, by no means all, old friendships fizzle out, because they get boring with the feeling that one has heard it all before. I don't believe I have the energy to develop this theme here, though it would be relevant here by contrast to the friendship with J.C.

Added after reading the obituary of J.C. in the *Daily Telegraph* (which Dr. Derus sent me). (a) It is a vivid reminder of the style in conventional obituaries. (b) By contrast, J.C.'s speech at his sister's 40th wedding anniversary quoted there as something to make one shudder is, of course, not, but may (have) evoke(d) second thoughts (in his sister Susanna, with whom he was on very good terms; I, too, found her very good company). He said on that occasion that she would have married the first man who came along. After 40 years marriage she probably knew better (than 40 years earlier).

19 January 2015 to Kenneth Derus. Chancellor sold 3000 books, a commercial success proportional to his interest in commerce, he spent the eve of his death at his brother's: watching his favourite child Anna, the actress, on TV. So much is surely clear from the obituary. Slowly I am getting over the shock of his death.

24 January 2015 to Kenneth Derus. As you probably realized, my letter to Faure complemented his report of the Memorial Service for Chancellor. Till I got your e-mail of 23.I.15 I did not know that one of his younger sisters had died when he was about 7, and she 5. I take it he did not tell you how he had arrived at the conclusion that this had an effect, good or bad, on his later development. *Remark* Even if he had told me about it I should not have asked for elaborating.

28 January 2015 to Kenneth Derus. Perhaps a stay at the hospital, where they discovered the weakness of my heart would help me, either by treating my heart or by infecting me with an antibiotic resistant disease and finishing me off. The present state of constant tiredness is intolerable. I am sure I have not answered all your questions, but I am too tired to concentrate adequately.

29 January 2015 to Kenneth Derus. Actually, I asked already Dr. Rosenlechner about Viagra. This seems to be well known, but he seemed to think it does not apply to me. Naturally, I'll ask at the hospital (of the Barmherzigen Brüder again, where the chief seems well informed about me). With luck, he knows how to strengthen the heart. I used to take digimerck, but they or at least Dr. Rosenlechner stopped.[39] If you want to know if I'm still alive, please, ask Spießberger.

30 January 2015 to Hubert Faure. From 2.2–13.2. I am due to go to the hospital that cleared up my double pneumonia. My heart is giving me trouble (being weak and making me continually tired). My private doctor knows no effective medicine.[40]

References

[Baaz and Wojtylak, 2008] M. Baaz and P. Wojtylak. Generalizing proofs in monadic languages. *Annals of Pure and Applied Logic*, 154:71–138, 2008.
[Kreisel, 1944a] G. Kreisel. A remark on the Schröder-Bernstein theorem. *Eureka*, 8:7, 1944.
[Kreisel, 1944b] G. Kreisel. On a geometric trifle. *Eureka*, 8:8–9, 1944.
[Kreisel, 1978] G. Kreisel. Review of *Wittgenstein's Lectures on the Foundations of Mathematics*, Cambridge, 1939. *Bulletin of the American Mathematical Society*, 84:79–90, 1978.
[Kreisel, 1984] G. Kreisel. Review of J.P. Burgess' "Dummett's case for intuitionism". *Zentralblatt für Mathematik*, Z0549.03006, 1984.

---

[39] Digitoxin was probably the only thing keeping him alive, but perhaps his heart could no longer tolerate it.

[40] Kreisel died in Salzburg on 1 March 2015 at 2:30a local time.

[Kreisel, 1987] G. Kreisel. Gödel's excursions into intuitionistic logic. In P. Weingartner and L. Schmetterer, editors, *Gödel Remembered*, pages 65–186. Bibliopolis, Naples, 1987.
[Kreisel, 1992] G. Kreisel. On the idea(l) of logical closure. *Annals of Pure and Applied Logic*, 56:19–41, 1992.
[Kreisel, 1993] G. Kreisel. Review of J. Hintikka's "Gödel's functional interpretation in wider perspective". *Zentralblatt für Mathematik*, Z0820.03001, 1993.
[Kreisel, 1998] G. Kreisel. Second thoughts around some of Gödel's writings: a non-academic option. *Synthese*, 114:99–160, 1998.
[Kreisel, 2011] G. Kreisel. Logical hygiene, foundations, and abstractions: diversity among aspects and options. In M. Baaz, C.H. Papadimitriou, H.W. Putnam, D.S. Scott, and C.L. Harper, Jr., editors, *Kurt Gödel and the Foundations of Mathematics*, pages 27–53. Cambridge University Press, New York, 2011.
[Rexroth, 1981] K. Rexroth. *Excerpts from a Life*. Conjunctions, Santa Barbara, 1981.
[Watson, 1938] A. Watson. Mathematics and its foundations. *Mind*, 47:440–451, 1938.
[Winters, 2000] Y. Winters. *The Selected Letters of Yvor Winters*. Ohio University Press, Athens, Ohio, 2000.